Toxic Contamination in Large Lakes

Volume IV   Prevention of Toxic Contamination in Large Lakes

Managing a Large Ecosystem for Sustainable Development

Edited by Norbert W. Schmidtke

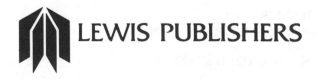 LEWIS PUBLISHERS

Library of Congress Cataloging-in-Publication Data

Prevention of toxic contamination in large lakes.

(Toxic contamination in large lakes ; v. 4)
Proceedings of a technical session of the World Conference on Large
Lakes, held May 18–21, 1986, Mackinac Island, Mich.
Includes bibliographies and index.
1. Water—Pollution—Environmental aspects—Congresses.
2. Environmental protection—Congresses. 3. Lake ecology—Congresses.
I. Schmidtke, N. W. II. World Conference on Large Lakes (1986 : Mack-
inac Island, Mich.) III. Series.
QH545.W3T67 vol. 4   363.7'394 s   87-36740
ISBN 0-87371-092-4   [363.7'3947]

LEWIS PUBLISHERS, INC.
121 South Main Street, Chelsea, Michigan 48118

PRINTED IN THE UNITED STATES OF AMERICA

PREFACE

The publication of **Toxic Contamination in Large Lakes**, a four-volume set of resource books, is the proceedings from the 1986 World Conference on Large Lakes. Governor James J. Blanchard and the State of Michigan hosted this international conference on Mackinac Island, May 18-21, 1986.

The proceedings represent research findings delivered at one of the first world conferences where government officials, policy makers, citizens, scientists, and business leaders joined together to explore solutions to toxic contamination in the large lakes.

Unique to the world conference was the emphasis on an integrated approach to examining toxic pollution in large lakes. The Conference covered not only scientific research, but also social and political concerns.

Over 500 world experts took this opportunity to share the latest research, management techniques, and political strategies for remedying large lakes from toxics.

The objectives at the Conference were:

1. Define the state of scientific knowledge with respect to toxic substances;

2. Identify successful, preventive, regulatory, and management alternatives;

3. Explore and encourage public education and involvement in lake management.

The primary concern of the conference participants was the fact that migrating toxic contaminants ignore political boundaries and can only be effectively controlled through international cooperation. Many of the discussions, as reflected in these proceedings, focussed on three key points concerning transboundary relationships in the management of the world's large lakes.

First, to resolve the problem of toxic contaminants requires a commitment from well-informed policy makers, scientists, citizens, and business leaders from all involved jurisdictions.

Second, jurisdictions must look beyond the political boundaries and study entire watersheds to establish prevention and control measures. New studies showing that airborne pollutants travel far greater distances than previously considered indicate the need for **ecosystem management** on a global basis.

Third, natural media - air, soil, and water - need to be managed by a cross-media approach.

These three points are well explored in the proceedings and deliberated upon in the conference synopsis that is featured in this publication.

Overall, the Conference represented a turning point for the future cooperation in protecting the world's large lakes. It brought the threat of toxic contamination to large lakes to the forefront of international environmental issues.

The documentation of this most important informal exchange will serve as a valuable reference to researchers, governments, business, and citizens around the world in their shared efforts to protect the earth's greatest freshwater resources.

William D. Marks
Chairman
World Conference on Large Lakes
Mackinac '86

ACKNOWLEDGMENTS

Special recognition is given to General Motors Corporation for underwriting the editing and copy preparation of the proceedings.

Other organizations that generously provided resources for the 1986 World Conference on Large Lakes and the publishing of the proceedings are: Allied-Signal, Inc.; Dow Chemical USA; Electronic Data Systems, Inc.; Hiram-Walker-Gooderham & Worts, Ltd.; The International Joint Commission; The Joyce Foundation; W.K. Kellogg Foundation; Mead Corporation; Michigan Department of Natural Resources; Michigan Department of Commerce; Michigan Sea Grant College Program; Charles Stewart Mott Foundation; Shell Western E & P; The Stroh Brewery Company; The Upjohn Company; and the United States Environmental Protection Agency-Great Lakes National Program Office.

The Shiga Prefecture of Japan is commended for originating the biennial world conference on large lakes in 1984. It was through the Shiga Prefecture's invitation to the State of Michigan to host the second World Conference on Large Lakes, that Mackinac '86 came to be.

Norbert W. Schmidtke is a consulting engineer and internationally recognized authority on wastewater treatment technology with 25 years experience in research, development, design, application and technology transfer. Prior to returning to consulting practice Dr. Schmidtke was director of Canada's Wastewater Technology Centre.

Dr. Schmidtke holds B.Sc. and M.Sc. degrees from the University of Alberta and a Ph.D. from the University of Waterloo. All degrees are in civil engineering. He is a registered professional and consulting engineer in the Province of Ontario, diplomate of the American Academy of Environmental Engineers and member of WPCF, PCAO, CAWPRC, IAWPRC, AEEP, CSCE and AIDIS. He is also associate editor (Environmental Division) of the Canadian Journal of Civil Engineering.

His areas of expertise range from industrial and municipal wastewater treatment process selection and analysis, to aeration technology, eutrophication, sludge and hazardous waste management. His special interest concerns process scale-up.

Dr. Schmidtke consults to industry, consultants, WHO and government organizations nationally and internationally. He has organized and is principal lecturer in many continuing education courses and custom-designed in-house training programs to industry, academia and governments worldwide. He was the 1981/82 national lecturer of the Canadian Society for Civil Engineering and recently represented CSCE on a lecture tour in the People's Republic of China. He is academically active through adjunct appointments in civil and chemical engineering.

Dr. Schmidtke has served as chairperson for many expert technical, advisory committees and Task Forces and recently chaired the Natural Sciences and Engineering Research Council's (NSERC) Grant Selection Committee for civil engineering.

Dr. Schmidtke has authored more than 130 technical papers and reports, and recently co-edited: 'Scale-Up of Water and Wastewater Treatment Processes'.

Four Volume Proceedings Author Index

Four Volume Proceedings Author Index

Contents

Contents

Contents

Contents

Contents

VOLUME III

SOURCES, FATE AND CONTROLS OF TOXIC CONTAMINANTS

Contents

Contents

VOLUME IV

PREVENTION OF TOXIC CONTAMINATION IN LARGE LAKES

Contents

Contents

WORLD CONFERENCE ON LARGE LAKES - MACKINAC '86

MAY 18-21, 1986

MACKINAC ISLAND, MICHIGAN, USA

Hosted by: JAMES J. BLANCHARD
                    GOVERNOR, STATE OF MICHIGAN

FOREWORD

The World Conference on Large Lakes brought together
scientists, policy makers, business and citizen leaders
to discuss issues associated with toxic contamination.
Significant issues confronting the quality of the
world's large lakes were chosen as conference themes
and explored during four concurrent sessions:

-Chronic Effects of Toxic Contaminants in Large Lakes

-Impact of Toxic Contaminants on Fisheries Management

-Sources, Fate and Controls of Toxic Contaminants

-Prevention of Toxic Contamination of Large Lakes

The complete record of the technical presentations is
contained in a 4 volume Proceedings set. In addition,
the co-conveners for each technical session prepared a
summary including findings and conclusions of their
session.   These summaries were designed to provide
conference participants and other interested parties
with a synopsis of the presentations and discussions at
Mackinac '86. The co-conveners then used their summary
to develop specific recommendations.

Each of the technical session's presentations, complete with session summary and recommendations, is presented in its separate Conference Proceedings volume. Volume 4 contains all session summaries and recommendations.

More than 500 people actively participated in a dynamic and productive exchange of ideas where almost 100 presentations were made. Conference planners and staff are grateful to all those who contributed their expertise and enthusiasm in this unique forum. We hope that the conference has been valuable to all who attended, and that relationships established here at Mackinac Island will support a growing effort to effectively prevent toxic contamination of large lake ecosystems worldwide.

A brief introduction to the topic of toxic contaminants in large lakes and conference summary follows.

INTRODUCTION

Large lakes contain a significant portion of the earth's water. For example, there are 253 large lakes greater than 500 km^2 which have a total surface area of 1,456,149 km^2 and an estimated volume of 202,000 km^3. Collectively, these 253 large lakes account for 93 and 88%, respectively, of the total surface area and volume of water held in all lakes of the world. These lakes, if only by virtue of the water they hold, constitute the world's most valuable resource. Large lakes fulfill key roles in the economy and overall well-being of humankind. They are used for drinking water, recreation, fishing, transportation, cooling water, and waste assimilation. Some large lakes, such as Lake Maracaibo in Venezuela are important sources of minerals and petroleum. Many large lakes, like the Great Lakes in North America, are particularly valuable as a source of drinking water and as a source of water for manufacturing. Large lakes, such as those in East Africa, also may be a life sustaining source of irrigation water.

Large lakes are also important for their ecology. The variety of large lake habitats result in a great species diversity. For example, Lake Baikal in the USSR, the world's deepest lake (maximum depth: 1,741 m), supports approximately 1,700 biological species, of which 1,200 are found nowhere else. Such species diversity is not only unparalleled, but of fundamental importance in maintaining the genetic pool and uniqueness of that aquatic ecosystem.

Many large lakes have suffered from overfishing, destruction of biological habitat (e.g. wetlands, spawning and nursery grounds for fishes), cultural eutrophication, or toxic substances pollution. In 1984, the first World Conference on Large Lakes Environment was held in Shiga, Japan. The theme of the 1984 conference was eutrophication. Michigan Governor James Blanchard accepted an invitation from Shiga (Michigan's sister state) to host the second conference focusing on toxic substances in large lakes.

Many toxic substances are highly persistent and can bioaccumulate in fishes and wildlife to levels which threaten aquatic ecosystens and human health. Large lakes, because of their long residence and flushing times (e.g. Lake Superior's flushing time is 181 years), are particularly susceptible to persistent toxic substances. The input of persistent toxic substances such as polychlorinated biphenyls (PCBs) and DDT has rapidly resulted in long-term, whole-lake problems in many large lakes.

Toxic substances can enter large lakes via manufacturing processes (e.g. industries), use (e.g. agriculture), and disposal practices (e.g. landfills). It is now clear that the atmosphere is a major pathway by which synthetic organic contaminants and trace metals enter large lakes. Long-range transport of toxic substances via the atmosphere is contributing to contamination of most large lakes throughout the world.

The 1986 World Conference on Large Lakes has brought together world experts on toxic substances in large lakes and key policy makers from jurisdictions which include large lakes. This conference has provided an opportunity for representatives of industrialized nations to share information on sources of toxic pollution, effects, and viability of control programs used to date. Representatives of less developed nations have had an opportunity to learn from the experiences of more developed nations. Policy makers have benefited from the opportunity to discuss transboundary problems, such as long-range transport of toxics through the atmosphere and multimedia problems, such as movement of toxic substances from landfills to lakes via groundwater seepage. The conference has also provided an avenue for concerned citizen leaders to share effective mechanisms for working with governments and business to control toxic substance pollution of large lakes.

CONFERENCE SUMMARY

The 1986 World Conference on Large Lakes has confirmed
the crucial need for the most basic of commodities -
information. Governments and public institutions have
a responsibility to communicate to citizens and the
media clear, concise, and accurate data concerning the
environmental health and well-being of the ecosystem.
For example, the public deserves more complete
information and criteria regarding fisheries closures
or advisories. Human exposure and chronic health
effects must be documented relative to long-term
exposure to multiple persistent toxic chemicals.

With regard to human health, the economics of the
fisheries, and the overall health of the ecosystem, the
solution is also simple - **stop the input of
contaminants**. The world is well aware that attention
should continue to be placed on the control of known
point sources. As an example, the report of the US-
Canadian International Joint Commission on the role of
non-point pollution in North America has served as a
strong basis for the development of new strategies to
cope with non-point source pollution. However, the
focus of most studies remains limited to conventional
pollutants. The next step for the world community is
to focus on the control of toxic substances for both
point and non-point sources and to monitor for emerging
pollutants.

Historically, scientific inquiry and public policy
development have been regarded as mutually exclusive
elements in the management of large lake systems. A
research brokerage function, designed to link science
and public policy is largely absent in large lake
management systems.

It is not enough that the scientific community
continues to recognize the interrelationships of our
environment through an **"Ecosystem Approach"**. The
regulatory community, through political processes, must
respond by implementing policy guided by a perspective
of our interrelated environment which extends beyond
national boundaries or environmental compartments and
must arrange their institutions accordingly. The world
community must adopt a philosophy of prevention of
toxic substance contamination rather than merely
reacting to environmental crises.

CHAPTER 1

CHRONIC EFFECTS OF TOXIC CONTAMINANTS IN LARGE LAKES

Technical Session Co-Conveners:

Douglas J. Hallett[1] and Donald C.McNaught[2]

[1] President, Eco-Logic, Inc., Acton, Ontario, Canada
[2] Director, Minnesota Sea Grant Institute, University
of Minnesota, St. Paul, MN 55108

SCALES OF SPACE AND TIME

Scientific evidence to date has shown that the
continued localized input of low levels of persistent
toxic chemicals into large lake ecosystems will produce
multiple effects in biota on a whole-lake basis. Long-
term buildups of multiple contaminants and subtle
effects have produced sudden catastrophic ecosystem
changes of threshold responses. Examples include the
viability of lake trout fry or herring gull
reproductive success or failure phenomena which have
shifted entirely within one year. Due to the size,
surface area, retention time, and dynamics of large
lake ecosystems, extremely long recovery periods are
necessary, if recovery is possible at all.   In
addition, the burden of proof for a cause/effect
relationship cannot be shown until the surprise,
catastrophe or irreversible ecosystem change has
occurred. For human populations, this is unacceptable.
Prevention, rather than reactive management, is
imperative.

ECONOMIC PLANNING

Toxic chemical management in contaminated large lake
ecosystems will require large, long-term budget
commitments. All measures, whether successful or not,
will require these expenditures. Therefore, all major
economic decisions should consider not only the
potential regulatory requirements in the foreseeable
future, but also the long-term ecosystem impact,
recovery time, and multiple generation effects. The
long reproductive time required in populations of fish,
wildlife, and particularly humans is of paramount
importance.

PLANNING THE ECOSYSTEM AND THE ECONOMY

Planning in business must allow for the recognition of
errors, audits, fast assessment of problems and
correction. So must planning for a sustainable
ecosystem. We must learn from the common toxic
chemical problems which are recurring on a global
basis: Minamata, Yusho, Seveso, Love Canal, Times
Beach, the Niagara River, Bhophal, the St. Clair River,
the Herring Gull, the Lake Trout, the Baltic Seal, the
Peregrine Falcon, the Bald Eagle.

While specific controls on primary sources, such as
severe restrictions on the use of certain hard
pesticides and mercury in North America, have met with
local success, secondary sources, such as atmospheric
transport from other countries, usage of compounds that
are not restricted, and historical in-situ sources,
such as contaminated harbour sediments and leaking
landfills have created a demonstratable plateau in
toxic chemical levels and the buildup of subtle chronic
effects within large lake ecosystems.

It is evident that insufficient information exists on
the production and usage of toxic chemicals in a global
regard and on a local scale. Further, the approach to
the hazard assessment of the multitude of chemicals
already present in the ecosystem is inconsistent and
inadequate. Insufficient information exists on the
exposure levels and chronic effects of environmentally-
derived toxic chemical residues on the health of human
populations. Minor chronic effects demonstratable on
environmental species today are, however, considered
major effects when manifested in humans.

RECOMMENDATIONS

1.A new thorough international effort is required now to eliminate the major retrievable sources of persistent toxic chemicals from not only large lake ecosystems, but the global ecosystem. A preventative approach must be taken rather than the present reactive approach which currently requires absolute proof of a cause/effect relationship, particularly proof of excessive risk to chronic human disease. Past experience demonstrates that multiple subtle effects at various trophic levels in an ecosystem due to multiple toxic causative factors, will build up over time resulting in irreversible population surprises, catastrophes or ecosystem crashes with corresponding detrimental changes to economy.

2.Human exposure and chronic health effects must be documented relative to long-term exposure to multiple persistent toxic chemicals in order to continue with regulations and decisions that require this data.

3.An inventory of sources for persistent toxic chemicals which are now prevalent within the ecosystem must be accurately obtained on a local, regional, national, and global basis.

4.Hazard assessment protocols must be developed and used cooperatively on an international basis to allow for worksharing and to provide internationally consistent legal decisions on toxic chemical control.

5.International control agreements are still required to:

a.eliminate hard pesticides such as aldrin/dieldrin, lindane, and toxaphene, and

b.provide for consistent and complete destruction of the major retrievable sources of persistent industrial chemicals.

6.Decisions on toxic chemical control must be based on a thorough, honest, and complete public disclosure of information.

CHAPTER 2

IMPACT OF TOXIC CONTAMINANTS ON FISHERIES MANAGEMENT

Technical Session Co-Conveners:

Niles Kevern[1] and H. Francis Henderson[2]

[1] Associate Director, Michigan Sea Grants College
Program, Michigan State University, East Lansing, MI
[2] Chief, FIRI, Food and Agriculture Organization of the
United Nations, Rome, Italy

As aquatic ecologists, limnologists, and fisheries
scientists and managers, we are often frustrated in our
efforts to convince the public and decision makers of
the importance of our large lakes or of our aquatic
resources in general. Narrowing the subject to
contaminants or toxic substances at this conference did
not alleviate this frustration. However, several
points seemed to be common among the presentations and
addressing these points as conclusions and translating
them into recommendations may help relieve the
frustrations and lead to more rapid progress in our
efforts to maintain the integrity and value of our
large lakes.

Most of us feel that we must use an ecosystem approach
in understanding and managing our large lakes. At the
same time, we realize that this is the most difficult
approach. It is the most difficult for the public to
understand and it is the most difficult to
institutionalize, as it requires a high degree of
integration and cooperation. So, what must we do to
achieve this ecosystem approach? We must of course,
continue our efforts to work in that direction and,
although time is precious, we must be patient and
persistent.

How do we gain public support for decisions that favor our large lakes? How do we impress the public with the value of these our lakes so that they, in turn, will support, even demand, our law makers to make the right decision? The key term is value. Our publics appreciate more readily those aspects that touch upon them directly. In this instance, they appreciate their water supply and they can appreciate their food, the fish they eat. They less readily understand constants governing solution or vaporization of chemicals, partitioning of organic compounds into fats and oils, or the phytoplankton and zooplankton interactions that are part of the aquatic foodweb. It is easier to understand that 24 million people rely on the Laurentian Great Lakes for their drinking water and that 13 million people rely on Lake Biwa in Japan for their drinking water. And that you get thirsty very quickly when your water supply is gone or that it would be extremely costly to remove all contaminants from seriously contaminated water.

It is easier to appreciate the lakes when we realize that the annual economic value of the salmonid fishery in the Great Lakes is 2-3 billion dollars and that the dollar value of freshwater pearl production in Lake Biwa is significant. We must ask our resource economist to continue to research and determine the worth of our large lakes. We must ask our social scientists to research and determine the non-market values of our lakes as perceived by the public. Hopefully we can realize a value high enough to convince decision makers to support additional research into subjects that require longer periods of time.

Several times in our session we noted that the public is often confused by our information on toxic substances in fish. We blamed the public media, but we also realized that we shared the blame. We have not pooled and coordinated our own information within or among our agencies. No wonder the public is confused. We tell them that eating fish is good to prevent heart disease and then issue health advisories about not eating fish because of contaminants. We need to do better in coordinating information among state, provincial and federal governments. In North America, the Great Lakes Fishery Commission (GLFC) is increasingly serving as a coordinating body among the state and provincial fishery management agencies. The GLFC and similar organizations elsewhere in the world need to place an even greater emphasis on this coordinating role. We must be more conscious of the need to present complete information to the media and when possible, to insist on more complete transferal to the public. The media must be aware of the very great role that they play in public education.

Part of our failure to get complete information to the public is because we are still missing information. We need to monitor our programs in contaminant control better. Much of our data to date are not standardized, leading to great variability and thus non-comparability. For monitoring purposes, we should standardize the species of fish (perhaps the lake trout), the size of the fish, the time of the year, the method of preparation, the analytical method, etc. We wouldn't remove all variability, but we could improve the comparability.

We also lack information on the real effects of contaminants on human health and on aquatic ecosystem health, thus we need long-term research on both aspects. Tolerance levels for contaminants in humans and in aquatic systems and action levels for fish must be based on continually improving data and on long-term, sub-lethal effects. Risk assessments are unreal unless such data are available. Realistic management of fish stocks is impossible without such data. For aquatic systems, we need microcosm or mesocosm approaches where the laboratory study becomes as natural as possible. Long-term population predictions for humans or fish are not realistic without data on carcinogen or mutagen effects.

In our session, we often observed a conflict between human health and economic values when establishing action levels of a contaminant for fish. This conflict, for lack of better information, leads to a judgmental action level. Ideally, with regard to human health, economics of the fishery and health of the ecosystem, the solution is really simple - stop the input of the contaminant. We watched the recovery of Lake Erie and heard about the beginning of the recovery of Lake Orta in Italy when nutrients or contaminant inputs to the lakes are stopped. We know that recovery can be good in most cases when we stop the input. We have done reasonably well at treating point sources to remove contaminants, but yet some point sources seem to elude us. We must bring increasing pressure to bear on known point sources and continue to identify unknown points. Increased effort is needed to reduce non-point input of known contaminants.

Again and again, speakers mentioned the vital role of public involvement. When the publics are convinced of the value of the resource, they will support the protection of that resource. We must convince them of the value of our large lakes.

RECOMMENDATIONS

1.More research is needed to determine the value, both economic and social, of our large lakes.

2.A responsible organization is needed to coordinate information for the public on toxics in fish.

3.More research is needed to establish realistic, long-term tolerance levels for fish health, aquatic ecosystem health, using micro-and mesocosms, and for the health aspects related to human consumption of fish.

4.Standardization of monitoring contaminants in fish is needed for comparing long-term trends and an early warning system.

5.Continuing efforts are needed to reduce or stop inputs of contaminants to large lakes from both point and non-point sources.

6.Develop a procedure to bring the public, possibly through advisory groups, into working groups with scientists and resource managers to address contaminant problems.

CHAPTER 3

SOURCES, FATE AND CONTROLS OF TOXIC CONTAMINANTS

Technical Session Co-Conveners:

Richard L. Thomas[1] and Wayland R. Swain[2]

[1] Director, International Joint Commission, Great Lakes
    Regional Office, Windsor, Ontario
[2] University of Amsterdam, Amsterdam, The Netherlands

INTRODUCTION

The origin of toxic chemical contaminants and their
ultimate environmental disposition are two major
factors in determining the necessity for their control.
There are over seven million known chemical compounds,
30,000 of which are in substantial commercial use.
Approximately 1,000 new chemicals are developed each
year. Over 1,000 chemicals are suspected carcinogens.
Some of these chemicals occur naturally, which further
illustrates the problem: manufactured chemicals are not
the only source of toxic substances.

As research in this area increases, scientists are
gaining a better perspective on the routes of toxic
contamination, yet the sources are so diverse and
numerous, it is very difficult to always draw a direct
relationship between sources and fate.

Toxic substances are transported via air, land, and
water routes and no single component can be separated
from the others in the ecosystem. The problem of
identifying the sources, fate and control of toxic
contaminants is one of the leading and most difficult
international issues confronting environmental managers
and decision makers today. Its resolution will require

an international effort in information exchange and
cooperation in mutual policy development.

It is now recognized that by the time that the sources
and fate of toxic substances have been identified it is
too late and the inevitable effects on the system must
be played out.   To overcome this, a philosophy of
prevention   is   needed   rather   than   to   depend   on
subsequent remedial actions.

SOURCES

The sources of toxic substances to the large lakes of
the world are many.   They include the direct inputs
from industrial, urban and mining activities, from
agricultural processes and pest control and from
associated waste disposal practices.   These loadings
are direct, variable and lake specific.   However, all
lakes are being perturbed by loadings of toxic
substances from the atmosphere.   Sources to the
atmosphere are however not well quantified.   Both
aquatic and atmospheric loadings do not respect
national boundaries.   Long range transport results in
the transboundary pollution of the world's large lakes.

Recommendations

    1.There is a need to more effectively manage
    and regulate toxic substances on a global
    perspective.   This approach must include
    developing nations.   International efforts
    must utilize scientifically valid techniques
    for the collection of compatible data.   This
    includes   the   use   of   internationally
    standardized protocols for the analysis cf
    toxic substances.   The data thus developed
    should be shared by nations so that a global
    response to the toxic exposure problem may be
    developed.

    2.Quantitative   and   qualitative   source
    information needs to be enhanced to provide
    adequate information for establishing mass
    balances as a means of establishing causes
    and   consequences.   Sources   include
    industries, municipalities, combined sewer
    overflows, agricultural and urban runoff, in-

place pollutants, contaminated groundwater, etc.

3.Mediation of environmental impact (e.g. toxicity and bioavailability) of in-place pollutants is needed.

4.It is recommended that developing nations be encouraged to use alternate approaches to persistent organochlorine compounds (e.g. DDT, Toxaphene) to obviate the buildup of these compounds in large lake systems. Emerging techniques in agriculture and pest control should be made available to developing nations as soon as possible.

FATE

The fate of toxic substances in large lake ecosystems is not completely understood, but is recognized as being extremely important because of contaminant uptake by aquatic organisms. The importance of uptake and release phenomena have only recently been recognized as major mechanisms for the movement of toxic substances through the aquatic environment. The velocities and movements of these water- and particulate-phase contaminants have been demonstrated to be extremely rapid, often resulting in lake-wide dispersion on a time scale of hours to days. Further, these mechanisms are responsible for the phenomena of biomagnification which results in substantial increases in contaminant burdens in succeedingly higher trophic levels in the food chain. Thus, food chain accumulation ultimately leads to human exposure, as man is the final predator in large lake ecosystems.

Recommendations

1.There is need for a process/mechanism oriented, multi-disciplinary, multinational, coordinated effort to understand the fate, transport and effects of toxic substances in large lakes. Existing historic, retrospective data sets on the North American Great Lakes, while unique in the world, have a substantial number of associated problems. In general, these data sets were assembled by individual researchers operating on various lakes, often in near isolation, and certainly using different methodologies, techniques and approaches. The data they assembled have been extremely useful in enabling us to achieve our present level of sophistication, but it is clear that future advances can only be made in the light of new knowledge.

Needed is a well-coordinated study combining the scientific expertise of participant nations to address the urgent unanswered questions with regard to transport, fate and effects of toxic substances in large lakes ecosystems. This study should be a carefully coordinated, long-term (5-7 year) program designed to assess the major mechanisms of transboundary distribution and dissemination of contaminants. Other nations are also encouraged to develop similar programs for large lakes of the world, and in particular to request international organizations to establish base-line studies in the large lakes of developing nations against which future developments and trends may be compared.

CONTROLS

A menu of strategies and techniques for the control of toxic substance release to the environment are available. They range from waste reduction at the source, to treatment, destruction and disposal. These strategies and techniques should be extended to developing nations.

Recommendations for Waste Reduction On-Site or Near Source

1.It is recommended that control be emphasized at or near the source through reuse, recycle, recovery, and waste exchange.

Recommendations for Waste Treatment, Destruction and Disposal On-Site or Near Source

1.It is recommended that hazardous waste management begin immediately.

2.It is recommended that an integrated approach to toxic substance control consisting of combinations of biological, physical, chemical and high-temperature thermal unit processes be implemented, recognizing opportunities for adding to existing in-place technology.

3.It is recommended that centralized hazardous waste treatment facilities for the small to medium sized generators be utilized.

4.It is recommended that innovative sludge management approaches and techniques be implemented.

5.It is recommended that existing or newly developed, low-cost, rapid screening techniques (e.g., bioassays) for assessing toxicity be used so that appropriate technical responses can be implemented before environmental damage occurs.

6.It is recommended that increased use of available process selection protocols for identification of the most cost-effective processes be implemented.

7.It is recommended that improved operation of existing hazardous waste treatment facilities which ensure that toxic compounds are not created (dioxins through incineration) be implemented. The importance of adequate operator training programs must be recognized and addressed.

8.It is recommended that effective technology transfer and continuing education be carried out in both developed and developing nations.

SUMMARY

It is not enough that the scientific community continues to recognize the interrelationships of our

environment through an "Ecosystem Approach". The
regulatory community, through political processes, must
respond by implementing policy guided by a perspective
of our interrelated environment which extends beyond
national boundaries or environmental compartments and
they must arrange their institutions accordingly.

CHAPTER 4

PREVENTION OF TOXIC CONTAMINATION OF LARGE LAKES:
Managing a Large Ecosystem for Sustainable Development

Technical Session Co-Conveners:

Jean Hennessey[1] and Ernst von Weizsaecker[2]

[1] Dartmouth College, Hanover, NH 03755
[2] Institute for European Environmental Policy
   5300 Bonn 1, Federal Republic of Germany

FINDINGS AND RECOMMENDATIONS

1.Historically, large lakes represent a vital economic
resource which has thus far has not been fully
appreciated.   The major uses, such as industrial,
transport, fisheries, domestic and recreational, are
determinant factors for the economic well being of
large adjoining regions.

2.There is a striking similarity between the socio-
economic and political characteristics of the world's
large lakes systems. This similarity is reflected in
the multi-jurisdictional or international institutional
arrangements developed to manage those systems.
Expanded information sharing through the recently
established International Lakes Committee and other
mechanisms may lead to innovative ideas for
strengthening institutional arrangements.

3.The management of this resource is frequently
rendered difficult by a multitude of political
boundaries, often including international boundaries.
Managing such transboundary environmental resources
represents a major challenge to existing political
institutions; therefore, a priority setting methodology

to guide research and management efforts is essential. In the Great Lakes Basin, the Science Advisory Board and the Council of Research Managers are designed to serve this function. Concerted action should be directed at developing the membership of the Science Advisory Board and the Council of Research Managers so that their recommendations are incorporated into the research priority setting mechanisms of the governments.

4.Historically, scientific inquiry and policy development have been regarded as mutually exclusive elements in the management of large lake systems. A research brokerage function, designed to link science and public policy, as well as provide a vehicle for applying it in a management context, is largely absent in large lakes management systems. It is recommended that research be undertaken into institutional alternatives for integrating scientific inquiry and public policy into a single system for water management in large lake systems.

5.International and inter-jurisdictional large lake systems are generally characterized by fragmented management frameworks based on political rather than ecosystem boundaries. Such fragmentation is reflected not only in institutional arrangements but individual jurisdictional laws, programs and policies. Further examination and research from socio-economic and legal-political perspectives must be undertaken to devise laws, policies and programs to protect the health and safety of citizens and which recognizes the primacy of the ecosystem boundaries while accommodating the realities of political jurisdiction.

6.Pollution is particularly threatening to large lakes which tend to become a "sink" for pollutants from all sources. Traditionally pollution sources have been characterized by environmental medium but it is becoming increasingly clear that the pathways of pollutants can be lead through several media and that controls consequently must take a cross-media approach.

7.Two themes were evident in the discussions of hazardous waste:

    i)prevention of hazardous waste is preferred to treating or disposing of such, and

    ii)existing hazardous wastes should be treated and disposed of immediately rather than stored because the social and economic costs of immediate disposition are less than

> the costs and long-term consequences associated with storage.

Research is needed to devise ways to prevent the formation of new hazardous wastes and to eliminate the hazards associated with those which already exist.

8. Intensive agricultural practices not only result in non-point pollution of air, land and water but also result in overproduction of foodstocks. Although economic incentives for implementation of practices which minimize or reduce agricultural and other non-point pollution do exist, such incentives are neither fully developed nor broadly applied. Policy and technical research is needed to develop and apply socially and economically feasible alternative agricultural practices.

9. The huge mass and diversity of environmental data overwhelm our ability to comprehend their meaning. Computer-based data and information management systems are needed to structure and organize this mass of information so that it becomes immediately useful to decision makers.

10. Many international commissions have been created to address concerns over shared regional resources. Their focus is on their own regional issues and between such commissions. A fostering of communication between international commissions addressing shared aquatic resources would provide exchange of information on their successes and failures. Valuable lessons and potential new initiatives could be learned from one another as was demonstrated at this conference by comparison of the management of the Baltic, Lake Geneva and the Great Lakes. Therefore, it is recommended that a conference of international commissions be convened to address institutional arrangements and the role of international commissions in addressing prevention and remediation of transboundary pollution.

11. Public awareness has been a vital motor of environmental policy which consequently has sought to use and develop appropriate means of public participation and information. Large lakes can only be successfully protected if citizens are aware of the importance of the issue and have the means of expressing their awareness. International private citizen groups have been able to overcome language and individual differences in providing demonstrable positive action on behalf of large lakes. Governments are encouraged to provide access to reliable information and to provide appropriate opportunities in

their decision making processes for participation by
the public.

12.Industrialized nations have gone through sequences
of naive pollution, awareness of problems, and problem
resolution. Developing countries appear to be readily
accepting the technology of industrialized countries
often without full recognition of the pollution
consequences. There is an urgent need for communication
between industrialized and developing countries on
pollution prevention so that the pollution experiences
of the more industrialized nations are not readily
repeated.

13.Historically large lakes have served as a critical
indicator of environmental quality. Research on the
Great Lakes has led to new understandings in such areas
as the management of fisheries, the bioaccumulation of
toxic substances and the role of phosphorus in
eutrophication. Nevertheless, in large measure, the
deployment of research has been crisis oriented. It is
a finding of this conference that basic, continuous and
integrative scientific research on large lakes should
be a high priority for governments.

14.The policy of the Great Lakes Water Quality
Agreement is reaffirmed, specifically the following
section:
     ..."The philosophy adopted for control of
     inputs of persistent toxic substances shall
     be zero discharge"...
It is further recommended that governments implement
programs advocated by the Royal Society of Canada and
the National Research Council of the United States in
their review of the Great Lakes Water Quality
Agreement.

15.Recognizing that the Pollution from Land Use
Activities Reference Group (PLUARG) study and report
conducted under the auspices of the International Joint
Commission was a landmark in the understanding of the
role of non-point or diffuse pollution in North America
and that it has served as a strong base for the
development of new strategies to cope with this
problem, it is proposed that a similar study be
undertaken for toxics. Such a study would assist the
Great Lakes Basin states and provinces in the
implementation of the precedent setting Great Lakes
Toxics Strategy which their governors and premiers will
commit to.   Therefore, it is recommended that the
governments of Canada and the United States request the
International Joint Commission to undertake a major
study on an integrated approach to the management of
toxic contaminants.

CHAPTER 5

AN ECONOMIC VIEW OF THE GREAT LAKES

Donna W. Wise

The Center for the Great Lakes, 435 N. Michigan
Suite 1733,Chicago, IL 60611, USA

INTRODUCTION

Throughout most of Great Lakes managements history,
economic development and environmental management
initiatives have been approached as being mutually
exclusive, viewed as occasionally complementary, but
more often conflicting. This philosophy has taken
Basin management in two directions now widely
recognized as undesirable. First, it fostered the
evolution of a Basin governance framework with limited
sensitivity to the "interconnectedness" of the
components of the Great Lakes system and the economic
and environmental attributes associated with them.
Second, in failing to recognize and capitalize on the
symbiotic nature of economic and environmental
considerations, this philosophy has historically
limited our opportunities to advance the health,
welfare and quality of life of the residents in the
Great Lakes Basin.

While in many respects our system of governance in the
region is a model for bi-national, interjurisdictional
co-operations and joint management, it has long
struggled to overcome a tendency toward fragmented
laws, programs and institutions. The origin of this
fragmentation is multi-dimensional, reflective of:

    1.a federalist system founded on political,
    as opposed to hydrologically-based
    jurisdictions,

2.the inherent tendency to "reduce" an often overwhelming management challenge to a more manageable series of compartmentalized tasks, and

3.an equal tendency to differentiate between the monetary values of a given resource and those values (e.g. aesthetic, quality of life) less given to quantification in economic terms.

This third dimension, while as pervasive as the two preceding it, is perhaps most amenable to change. The distinction between economic development and environmental goals is as artificial as the "dotted line" on the map that "separates" the waters of the Great Lakes and allocates them among the various jurisdictions. Indeed, the inseparability of economic and environmental considerations in pursuing goals for the Great Lakes region is more than simply fact. It is a fundamental management principle that cannot be ignored.

This notion sets the theme for subsequent discussion that documents the importance of the Great Lakes to the region's economy. It highlights success stories, obstacles and challenges and suggests a course or set of guidelines for future action in addressing goals for the Great Lakes region.

TOWARD A "NEW" MANAGEMENT PHILOSOPHY

A philosophy which recognizes the inseparability of economic and environmental considerations in Great Lakes management is now emerging among the region's leadership. This philosophy is actually an extension of the ecosystem management concept that has been developing for almost a century and was formally recognized in this region through the 1978 United States-Canada Great Lakes Water Quality Agreement (IJC 1978). The "ecosystem approach" reflected in that Agreement views man as part of the natural system rather than separate from it; as a participant rather than spectator. It recognizes that the complex interactions within the Great Lakes watershed or other large lakes system are more than physical, biological and chemical processes. They include man's political, social and economic attributes as well. Therefore, recognizing the necessity of linking economic considerations in Great Lakes management is a critical element in continuing efforts to develop and apply the ecosystem management concept in the Great Lakes or any other large lakes setting.

THE ECONOMIC SIGNIFICANCE OF THE GREAT LAKES - A LONG KEPT "SECRET" RECEIVES UNPRECEDENTED ATTENTION

To borrow a metaphor, our historical failure to perceive the economic significance of the Great Lakes has been a case of not seeing the forest for the trees. The contribution of the lakes to the region's economic development has been pervasive:

> -The role of the lakes in this regard dates back centuries. In fact, the first recorded incident of modification of the Great Lakes system dates back to 1797, when the North American Fur Company installed a lock at Sault Ste. Marie to improve transportation of furs for commercial sale.

> -The mere physical presence and geographic configuration of the systems was, and continues to be a determinant of locational decisions for business and industry. In fact, virtually every major city in the Great Lakes Basin is located on the shores of the Great Lakes or a tributary to them.

> -Much of the early economic activity during settlement of the region was directly attributable to the resource exploitation potential and availability of water-based transport. Even today, with a diversified regional economy, the Basin's water resources play an important role in attracting and retaining the industrial base.

Over time perhaps, we began to take the lakes, and their contribution to our economic well being, for granted. Like tap water and electricity in our homes, we came to view the lakes as unfailingly available, reliable and inexhaustible. We began losing our appreciation for them and with that loss, a sense of stewardship responsibility for taking care of them.

There is evidence, however, that this trend has been halted. Only recently have a series of events triggered a new found awareness and appreciation of the economic significance of the lakes. For example:

-The region's deep recession earlier this decade prompted Great Lakes governors and premiers to take stock of their economic assets. In so doing, they found the key to their economic future in the lakes themselves, through their protection and development for sustainable use. The signing of the Great Lakes Charter in 1985 by the region's governors and premiers was a consequence (Council of Great Lakes Governors 1985).

-The continued depletion of the Ogallala aquifer and emerging water shortages in southwestern regions of the United States has rekindled interest in some sectors for large-scale diversion schemes to mitigate shortages (Donahue et al. 1986). The estimated economic impacts on the region from reduced water levels and flows have been cause for alarm (IJC 1985).

-Toxic contamination incidents are increasingly recognized as a threat to water quality and the implications for acquisition of alternate supplies are tremendous. The economic hardships caused by abandonment of municipal and domestic wells due to groundwater pollution are now documented throughout the Basin.

These and other developments have brought economic considerations to the forefront of regional water policy discussion. Further, they have reminded all users of the resource of three fundamental yet often overlooked facts. Very simply:

1. water is an economic asset and has monetary value,

2. clean water is more valuable than dirty water, and

3. the Great Lakes have a large yet finite supply of water.

These facts address the very basic principles of economics, the concepts of supply, demand and product quality. From an economic argument alone, stewardship of the resource for long-term, sustainable use is good business. When considered in the context of environmental quality, human health and other difficult to quantify terms, the argument for careful stewardship is an even more compelling one.

A SURVEY OF GREAT LAKES WATER USE AND ECONOMIC BENEFITS

In building a case for increased recognition of the
economic significance of the Great Lakes we cannot rely
on concepts or philosophies alone.   Hard facts and
examples are needed.   The Council of Great Lakes
Governors recognized this when, in 1983, it asked the
Center for the Great Lakes to analyze the relationship
between the availability of Great Lakes water and
future economic growth in the region.   Hundreds of
personal interviews and report analyses were carried
out to:

   -document the ways water is used by different
   economic sectors in the region;

   -gather estimates of the amount of money
   generated for the regional economy or
   particular water dependent sectors of that
   economy; and

   -isolate examples of innovative marketing and
   promotional strategies that focus on the
   region's water resources.

This information was reported in August, 1984 (The
Centre for the Great Lakes 1984).

Before reviewing selected areas of Great Lakes water
use and economic benefits it is appropriate to place
the resource in context with the population it serves.
In so doing, it is difficult to avoid the frequent and
almost tiring use of superlatives.   It is an ecological
system of virtually unfathomable expanse and
corresponding complexity, yet a delicately balanced
one, susceptible to the subtlest of environmental
stresses.   As an expansive and intensively used
freshwater system, the lakes enjoy a global prominence.
The system contains 2.4 x 10^{12} m^3 of fresh surface
water; a full twenty percent of the world's supply and
over ninety percent of the United States supply (IJC
1985).   Its component parts, the five Great Lakes, are
among the fifteen largest freshwater lakes in the
world.   Collectively, the lakes and their connecting
channels comprise a system with a surface area of more
than 95,000 square miles a drainage area of well over a
quarter million square miles (IJC 1985).   As both an
international border and shared resource, the system
extends some 2,400 miles from its westernmost shores to
the Atlantic, a distance comparable to a trans-Atlantic
crossing from the east coast of the United States to
Europe.   As the continent's "fifth seacoast", the
system provides many thousand miles of coastline.

A review of the Basin's socio-economic statistics is also of use in providing a context or perspective for examining the interrelationship between economic and environmental considerations. Consider the following:

   -within the Great Lakes Basin reside 20% of the entire U.S. population and 60% of the Canadian population; over 40 million residents.

   -over 25 million residents rely on the surface waters of the Great Lakes for their drinking water.

   -one-fifth of all U.S. manufacturing is located along the Great Lakes shoreline. About half of Canadian manufacturing is located within the Great Lakes Basin in Ontario. All such activity is dependent upon access to abundant and, in many cases, high quality water supplies (The Centre for the Great Lakes 1984).

   -more than 145 billion liters of Great Lakes water are withdrawn daily to supply the domestic, commercial, industrial and agricultural needs of the basin's residents (Great Lakes Basin Commission 1979).

   -the Lakes support a recreation/tourism industry that is worth billions of dollars to the Michigan economy alone.

   -the Great Lakes/St. Lawrence Seaway System adds $3 billion to the Great Lakes economy every year, and in so doing, provides a vital link between the nation's agricultural and industrial heartland and ports throughout the world (St. Lawrence Seaway Development Corporation 1984a).

Clearly, the abundance and quality of the resource shapes the environmental and socio-economic fabric of the region. Just imagine the converse; the ramifications of Basin-wide water scarcity and widespread contamination. To avoid such an eventuality, each of us must recognize our vested interest in the quality and availability of the Great Lakes resource.

To inventory the uses of the Great Lakes is to inventory our livelihoods and our pleasures. From hydropower and hydroplanes, from industrial cooling water to drinking water, our dependence on the lakes is indeed pervasive. Consider just a few uses:

1. <u>Domestic Water Supply</u>.   Every day, well over 11
billion liters of water are withdrawn from the lakes to
satisfy the needs of the Basin's 40 million residents.
(Great Lakes Basin Commission 1979). Well over half of
these individuals rely on the surface waters of the
Great Lakes for their drinking water.

2. <u>Industry</u>.   The Great Lakes comprise the industrial
heartland of North America.   Over 50% of U.S. heavy
industry and 70% of its domestic steel is produced in
the Great Lakes region (Great Lakes Economic Policies
Council 1983). Over 51 billion liters of water per day
are withdrawn from the lakes for commercial, industrial
and   manufacturing   purposes   (Great   Lakes   Basin
Commission 1978).

3. <u>Transportation</u>.   In more than 25 years since the
opening of the St. Lawrence Seaway, over a billion
tonnes of cargo, with a value of more than $200
billion, have moved through the Seaway to and from
ports in North America, Europe, Asia, Africa and the
Middle East (St. Lawrence Seaway Development Commission
1984b). Research by the Center for the Great Lakes has
shown that every time a ship calls at a Great Lakes
port for an average load of cargo, almost a quarter of
a million dollars is generated for the local economy
(Centre for the Great Lakes 1985).

4. <u>Recreation</u>.   Great Lakes-based recreation adds up to
a multi-billion dollar business.   For example, over
one-third of all registered boats in the United States
are located in the Great Lakes states (National Marine
Manufacturers Association 1985).   Six of these states
are in the top ten nationally.   Shoreline parks along
the Great Lakes in the United States and Canada
attracted 63 million visitors in 1983.   Over 3.7
billion dollars were added to the regional economy as a
result (National Park Service 1984).

5. <u>Power Production</u>.   In Lakes Ontario and Erie, over 73
billion liters of water per day flow through hydropower
facilities.   In 1984 Great Lakes hydropower produced
almost 47 billion kilowatt hours of electricity in the
U.S. and Canada (Patterson et al. 1985).

6. <u>Agriculture</u>. The climate of the Great Lakes region
is tempered by the lakes, resulting in an extended
growing season and providing a readily available
irrigation supply. Forty percent of the Basin's land
area in the United States is in agricultural
production. These states produce over one-fifth of the
value of total U.S. agricultural exports, and along

with adjacent states to the west, produce almost 75% of
the nation's grain crop.   Agricultural production in
the Great Lakes provinces is equally significant.

7.Shoreline Revitalization and Redevelopment. A recent
survey by The Center for the Great Lakes found that 43
of   64   shoreline   communities   contacted   have   new
waterfront   developments   in   operation,   under
construction or in advanced planning stages (The Center
for the Great Lakes 1986).   In Detroit, for example, a
$3 million investment in a riverfront park has already
generated $200   million   in   waterfront   business   and
residential   development.   Further,   in   Toledo   a $14.5
million investment in the Portside development created
4,000 new jobs and generated $10 million last year in
tax revenues for the city.

These are clearly many uses for the Great Lakes that
translate   into   monetary   value.   But   perhaps   most
importantly,   one   must   consider   the   aesthetic   and
ecological value of the Great Lakes.   Values which are
not easily quantified but nonetheless substantial.

While unquestionably impressive, these statistics are
but   a   shadow   of   the   full   potential   that   might   be
realized   by   managing   the   resource   for   long-term
sustainable   use   in   an   environmentally   responsible
manner.   The   success   stories   are   there: burgeoning
waterfront   development   activity,   a   revitalized   sport
fishery and water based tourism industry, advances in
the   control   of   municipal   point   source   pollution,
growing   levels   of   interstate   and   international   cooperation, and others.   The challenge before us is to
fully recognize and work toward this potential.

MEETING THE CHALLENGE

In order to safeguard and build on our accomplishments
in the region, several principles must be embraced and
reflected in management efforts.

The term management is used here in its broadest sense:
the political, social and economic measures taken at
both the   public   and   private   sector   levels   toward   a
common   regional   goal.   Four   such   principles   are
presented below.   All are fundamental and perhaps self
evident,   yet   long   overlooked.   Each   rests   on   the
fundamental premise of this paper - the inseparability
of economic and environmental considerations in the use
and management of the Great Lakes:

1.Our public institutions and private businesses must reject parochial tendencies to either isolate or merely balance environmental and economic considerations in resource management efforts. Integration is the key: looking at water quality and recognizing its economic implications for waterfront development, or looking at the revitalization of water-based industry and recognizing its contribution to the tax base and subsequently water management programs.

2.We must recognize that stewardship responsibilities are not exclusively public sector responsibilities. We all have a vested interest in the future of the resource. Indeed, we should certainly care about something we drink, swim in, take food from, look at and live near. We must work toward the creation of our own "research triangle" here in the Great Lakes - a cooperative venture among government, academia and industry - to move our region forward.

3.In that same vein, we must explore and strengthen the linkages between these sectors and among our existing public institutions. In a sense, the collective management effort in the region is something of an imperfect puzzle where most of the pieces exist but do not yet properly fit.

4.Finally, we must take our message to the region's citizens. The Great Lakes are more than merely a tourist attraction or source of drinking water. They are not only a natural asset, but a necessity and a determinant of our environmental quality, our economic well-being and our quality of life.

These observations are not unique to the Great Lakes Basin, but have applications to any large lakes setting.

In closing it should be emphasized that our true challenge is not simply the recognition of the inseparability of economic and environmental considerations in large lakes management. The challenge lies in translating that recognition into programs for managing our lakes without compromising their environmental integrity, either now or in the future.

ACKNOWLEDGEMENT

The author acknowledges the assistance of Michael J. Donahue, Director, Chicago Office and Head of Research, The Centre for the Great Lakes, in the preparation of this paper.

REFERENCES

Center For The Great Lakes 1986. Water works! A survey of Great Lakes waterfront development on U.S. and Canadian shores. Chicago, Illinois and Toronto, Ontario.

Center For The Great Lakes 1985. Moving government goods on the Great Lakes: What an increase would mean in dollars and jobs for the region. Chicago, Illinois and Toronto, Ontario.

Center For The Great Lakes 1984. The lake effect: Impact of the Great Lakes on the region's economy. Chicago. Illinois.

Council of Great Lakes Governors 1985. Great Lakes Charter. In: Water Diversion and Great Lakes Institutions. January 1985. Madison, Wisconsin.

Donahue, M. J., A. Bixby and D. Siebert 1985. Great Lakes diversion and consumptive use - the issue in perspective. In: Seminar Papers of the Great Lakes Legal seminar: Diversion and Consumptive Use. Journal of International Law, Case Western Reserve University, pp. 19-48. Cleveland, Ohio.

Great Lakes Basin Commission 1979. Great Lakes Fact Sheets. Ann Arbor, Michigan.

Great Lakes Economic Policies Council 1983. A profile of the Great Lakes states. Cleveland, Ohio.

International Joint Commission 1985. Great Lakes diversions and consumptive uses. Report to the Governments of the United States and Canada under the 1977 Reference. January, 1985. Windsor, Ontario.

National Marine Manufacturers Association 1985. Boating Statistics - 1985. Chicago, Illinois.

National Park Service, Midwest Regional Office. Omaha, Nebraska and interviews with the Great Lakes states and provincial agencies conducted by The Center for the Great Lakes, June 1984.

Patterson, R., Edison Hydro; J. Bray, Sault Area Engineer, U.S. Army Corps of Engineers, Buffalo District and J. Whiteway, Research Analyst, Ontario, Hydro. Interviews with The Center for the Great Lakes, March 1986.

St. Lawrence Seaway Development Corporation 1984a. Internal Corporate Communication, July.

St. Lawrence Seaway Development Corporation 1984b. The Seaway. In: Seaway Review, July-August 1984.

CHAPTER 6

THE HISTORIC ROLE OF A LARGE LAKE IN JAPAN -
THE CASE OF LAKE BIWA

Kira Tatuo

Lake Biwa Research Institute, Otsu, Shiga 520, Japan

INTRODUCTION

Lake Biwa is the only inland water body in Japan that deserves the name of a "large lake" according to the definition given in the prospectus of this conference. It has a surface area of 674 km^2 and a maximum depth of 104 m.   Located close to the oldest capitals and big cities such as Nara, Kyoto and Osaka, it has played a significant role throughout Japan's history.

The lake is located nearly at the center of the Main Island of Japan in Shiga Prefecture, one of the 47 autonomous provinces of the country.   Its' water drains in a southwesterly direction to the Osaka Bay via the Yodo River.   The river serves as the effective stabilizer of stream flux.   Its catchment area almost overlaps with the administrative area of Shiga Prefecture, which government sponsored the first world lake environment conference in 1984.

ANCIENT TIMES

The main functions of rivers and lakes for pre-modern societies were to supply fishery products and serve as inland traffic routes.   In a country like Japan where rivers are short and mostly torrential (Figure 1), lakes were particularly important as highways for boat transportation and traffic.   There is archeological evidence that neolithic fishery settlements had already existed as early as 7,000-8,000 years ago.   Canoes were used on the lake at least 3,000 years ago (Table 1).

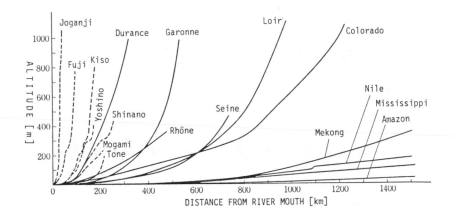

Figure 1.   Comparison of profiles between Japanese (broken lines) and continental rivers (solid lines).   From: Sakaguchi, Y., Takahashi, Y. and Omori, H. (1986): Rivers in Japan (nature in Japan, Vol 3), Iwanami Shoten, Tokyo, 248 p. (in Japanese).

Toward the end of the prehistoric period, there were a considerable number of local political powers on the Japanese Archipelago.   Their contacts with Chinese dynasties at the beginning of the 1st century A.D. had already been recorded in the early Chinese history books.   Later in the 5th to 6th century, the central district of the Main Island including Nara, Osaka, Kyoto and Shiga areas gradually took the leadership over the other parts of Japan until the first unified empire centering on Nara Province was formed in the middle of the 7th century.   This occurred after years of scramble among local political powers and, to a certain extent, against the strong pressure from continental kingdoms.

The earliest Japanese capital, Nara, was constructed in 710 A.D. and modelled after the Chinese metropolis of Chang-an.   It was also the center for newly introduced Budhism where big temples like the well known Temple of

TABLE 1   CONVENTIONAL DIVISION OF PERIODS IN JAPANESE HISTORY

[Archeological period]	[Period]	[Political system]	[Capital]
Paleolithic			
—10,000 B.C. (Appearance of the earliest pottery)			
Neolithic — Period of Jomon Pottery			
—300 B.C. (Introduction of rice culture)			
Period of Yayoi Pottery			
300 A.D.			
Period of Ancient Burial Mounds		Progress of unified state formation	
600 A.D.			
Historical Period	Asuka Period	Emperor's government born in Nara district	
710 A.D.	Nara Period	Emperor's government	Nara
784 A.D.	Heian Period	Emperor's government	Kyoto
1,185 A.D.	Kamakura Period	Hojo Shogunate(feudal)	Kyoto* Kamakura**
1,336 A.D.	Muromachi Period	Ashikaga Shogunate(feudal) Civil War 1,467-1,573 A.D.	Kyoto
1,568 A.D.	Azuchi-Momoyama Period	Civil War ceased under the Leadership of Oda-Nobunaga and Toyotomi-Hideyoshi	Kyoto
1,598 A.D.	Edo Period	Tokugawa Shogunate(feudal)	Kyoto* Edo(Tokyo)**
1,867 A.D.	Modern Period	Constitutional government	Tokyo

*Ritual capital.   **Political capital.

Huge Statue of Buddha were built. A historical record
tells of conifer timbers for the temple construction
being cut on Mt. Tanakami close to the southern tip of
Lake Biwa and transported on the Yodo River System to
Nara. In the 8th century the Nara government owned
three timber production centers within Lake Biwa's
catchment area, which was formerly called the Province
of Omi or Freshwater Sea. Omi has been known since
ancient times for its rich production of rice, iron,
timber and fish. The coastal plain surrounding Lake
Biwa was one of the districts where wet paddy
cultivation was first introduced from continental Asia
about three centuries B.C.. Omi became the largest
center of rice production in ancient and medieval times
as shown by statistical records of the 10th century
(Figure 2). The knowledge of iron smelting and
probably of charcoal making, which is indispensable for
metal smelting, was also introduced from continental
Asia across the Japan Sea to the Province. This
contributed greatly to the wealth of the ancient
Japanese Empire. Local powers which controlled the
iron production in Omi Province are said to have played
an important role in the dawn of Japan's history
(Figure 3).

The Nara Period terminated at the end of the 8th
century when a new capital was founded at Kyoto in 794
A.D.. Since then, Kyoto has been the Emperor's capital
for more than 1,000 years, until 1868. The city lost
the function of political center after the
establishment of Tokugawa Shogunate in Edo (old name of
Tokyo) in 1603. Since Kyoto is only ten kilometers
from the southeastern shore of Lake Biwa, it also
depended on the supply of primary industry products
from the lake and its drainage basin, just as Nara did.
However, Kyoto had grown into a city of more than half
a million inhabitants at the beginning of the Edo
Period. It was comparable in size to the biggest
cities of the contemporary world, like Paris and
London. Therefore, it required many more products for
its sustenance than just the supply from Omi and the
other neighboring provinces.

Overland transportation by means of carts and wagons
was not well developed in Japan. This was mainly due
to topographic reasons. Boat transportation on Lake
Biwa therefore served as the lifeline for Kyoto. The
transport of rice, seafoods and other products from the
northern provinces along the coast of the Japan Sea
across Lake Biwa had been maintained since the earliest
stage of the history of Kyoto. Those products brought
to a few sea-ports to the north of Lake Biwa were then
carried on horseback to the northernmost shores of the
lake for further southward boat transport (Figure 4).

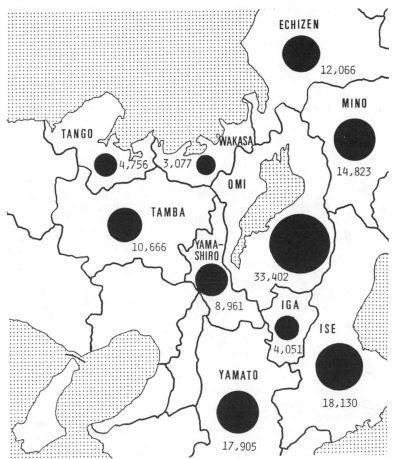

Figure 2.   Distribution of permanent rice fields in the
            provinces around Lake Biwa in the 10th
            century.   Numbers indicate the total rice
            field area per province in units of Cho
            (equals ha).

FEUDAL TIMES

With the urban growth of Kyoto more and more ports
became available on the lake shores for trans-lake
traffic.   Later, Osaka, a sea-port on the mouth of the
Yodo River developed rapidly into a big commercial
city.   This development occurred since the end of the
16th century.

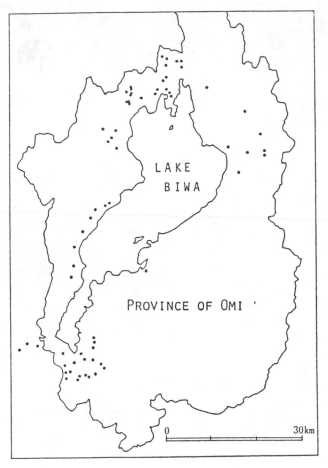

Figure 3.   Remains of ancient (about 5th-10th century)
            iron smelting in Shiga prefecture.   From:
            Ishihara, S. and Maruyama, R. : Korean
            Influences on Ancient Omi, Jimbutsu-Orai-
            Sha, Tokyo, 239p. (in Japanese).

Throughout medieval ages, the traffic and fishery
activities on Lake Biwa had not been controlled by the
central government but by private organizations.   For
instance, Katata, a village on the western shore north
of Otsu, controlled the whole lake surface.   It was
then a kind of autonomous community consisting of
samurai, fishermen and farmers which monopolized the
rights of off-shore fishery, pilot service and boat
building.   With backing of their armed forces they even
collected transit tax from lake boats.   Later, however,

the use of the lake and its resources tended to be more centralized. This occurred after the strong feudal government of Tokugawa Shogunate in Edo had been well established. The shores though, belonged to some fifty feudal lords.

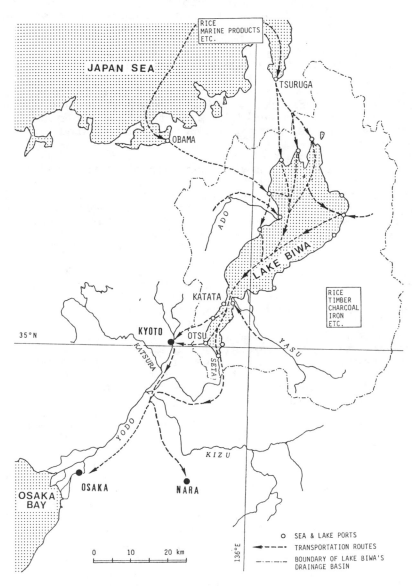

Figure 4. Boat transportation routes on and around Lake Biwa in ancient and medieval ages.

It is estimated that about 1,300 boats were engaged in transport service on Lake Biwa in the 17th century. Otsu (lit. big port), near the southern end of the lake, the present capital of Shiga Prefecture, began to thrive as the final landing place of southbound boat cargo. The beach of the city was lined with white-washed warehouses for rice storage. A ferry service was operated between Otsu and Yabase across the southern part of Lake Biwa as a shortcut to the highway from Kyoto to Edo (Figure 5). With the help of lighthouses navigation used to continue even after sunset. These lighthouses were actually big stone lanterns of traditional oriental style built mostly by the local communities. The one standing in front of the Lake Biwa Research Institute today (Figure 6), was built in 1845. Carved in its surface are the names of many traders and other persons who contributed to its construction.

Flooding caused by heavy summer rainfall and high water level frequently threatened the shores of Lake Biwa, where settlements and paddy fields had already advanced close to the beach line during the Edo Period. The flooding was essentially a natural disaster due to concentrated rainfall and limited draining capacity of the lake's single outlet the Seta River. In part, it was also caused by the destruction of hill slopes and their vegetation. To the southeast of the lake on the left bank of the Seta, the hills are called the Tanakami Mountains. Here the timber for the construction of the ancient city of Nara was harvested. The hills consist of deeply weathered, fragile granite, so that lumbering as well the continuous harvesting of firewood and undergrowth by farmers in later years, made the hill slopes almost devoid of woody vegetation. In places the hills were almost bare. All this resulted in accelerated siltation of the Seta riverbed. This in turn resulted in poorer drainage by the river and increased the danger of flooding. During the Edo Period, the Shogun's government was forced to permit the dredging of the Seta. This was in response to the repeated appeal from the lakeshore villages, even though the government was not willing to do so for military reasons and opposition by the residents of Osaka and other downstream areas. The dredging operations were, however, not very effective. This was probably the only environmental issue in Lake Biwa before the westernization of Japan.

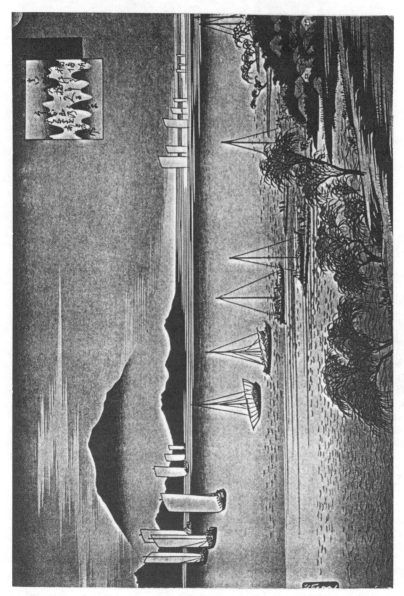

Figure 5. Ferry boats from Otsu arriving at the port of Yabase.
A woodcut picture by Ano-Hiroshige (1797–1858).

Figure 6.   A stone lighthouse built on a beach at Otsu
             in 1845.   Lake Biwa Research Institute is
             in the background.

MODERN TIMES

At the end of the 1867 revolution (or the so-called
Meiji Restoration), the Tokugawa Shogunate with its
feudal system was abolished and replaced with a new
government based on western style bureaucracy.   The
leading policy of the new government was to catch up
with the western countries as soon as possible.   The
whole country was reorganized as an assemblage of
prefectures.   Lake and sea surfaces were designated as
public property.   It is worthy to note that besides
national defence and education, highest priority in the
area of public investments was given to railway
construction and flood control.

The Tanakami Mountains became the first test case for
erosion control technology introduced by Dutch
engineers.   In 1878 the first flood control dam in
Japan was constructed as part of the flood control
program for the whole Yodo River system.   Frequent
flood disasters (1886, 1889 and 1896) demonstrated the
urgency for additional flood control measures for the
river.   The unparalleled flood in 1896, which caused a
maximum rise of water level of 3.8 m and submerged some

58,000 houses around the lake for months, forced the government to initiate a second improvement project. It was completed in 1906 with the construction of the Seta River Barrage for stream discharge and lake water level control (Figure 7). Extensive dredging of the river bed increased the Seta River's draining capacity by several times. The watershed of the Yodo, including Lake Biwa, was the target for flood control measures by the new government. Improvement work has continued until today so as to ensure a degree of safety for a very large population in the downstream districts.

Figure 7. Seta river barrage. First built in 1906 and completely reconstructed in 1961 for controlling the water level in Lake Biwa and the outflow of its water to the Seta River.

Similar initiatives in civil engineering were also taken by a pioneering attempt to utilize Lake Biwa's water resource. As a result of the above-mentioned government policy, inland water transportation became promptly displaced by newly built railways. The trunk railroad line called Tokaido Line that connected the two civilized areas of Japan, Tokyo/Yokohama and Kyoto/Osaka/Kobe, was opened in 1889. The line passed along the southeastern shores of Lake Biwa. This, and the formation of a railway network that followed, contributed to the rapid decline of waterborne transport on the lake, to the extent that the lake was eventually regarded as an obstacle for land transportation. On the other hand, the potential value

of Lake Biwa as a large water resource was recognized
at the same time.   This was attributed to an ingenious
young engineer.   At the request of the governor of
Kyoto Prefecture, Tanabe-Sakuo (1861-1944) designed and
supervised the construction of a canal that conducts
water from Lake Biwa to Kyoto city.   The Lake Biwa
Canal (Figure 8) was completed in 1890, only one year
after the opening of the Tokaido railway line.   This
illustrates quite dramatically the change in the role
of the lake from pre-modern to modern times.

Figure 8.   Lake Biwa canal completed in 1890.

Originally, the canal was a multi-purpose canal.   It
was intended to serve as a source of water for domestic
use,   irrigation,   replenishing   urban   streams   for
hydroelectric power generation and transportation by
boats.   A hydroelectric plant started operation in
Kyoto in 1981.   It is one of the earliest recorded
cases of hydroelectric power supply.   The city of
Wisconsin,   U.S.A.   initiated   hydroelectric   power
generation nine years earlier.   Kyoto became the first
city in Japan with electric streetcars (1885).   The
city was then losing its prosperity due to the transfer
of the national capital to Tokyo as the result of the
Meiji Restoration.   However, the impact of Lake Biwa
Canal was such that it could successfully stimulate
Kyoto to be revived as a modern city.   The need for an
assured supply of electricity and city water grew so

steadily that a second canal and a filtration plant for the city waterworks were constructed between 1908 and 1912.

Evidently, Lake Biwa contributed considerably to the initiation of technological, industrial and urban developments in modern Japan. Not to mention the Osaka case which, without the ample supply of lake water via the Yodo River, could never have grown into the largest industrial and commercial city in prewar Japan.

At present, approximately 13 million residents in Shiga and the other three downstream prefectures, including such large cities as Osaka, Kyoto, Kobe, etc., depend either directly or indirectly on the water supply from Lake Biwa and its downstream rivers. Osaka is by far the greatest consumer (Figure 9). According to a government estimate, an additional supply of about 67 m^3/sec will be needed by 1990 for domestic, industrial and agricultural purposes. To cover this increasing demand, a new water resource development project, called the Lake Biwa Comprehensive Development Project, was initiated in 1972. It is expected to continue until 1991. The objective is to increase the rate of lake water outflow at the Seta Barrage by 40 m^3/sec over the present rate (averaging approximately 150 m^3/sec between 1901 and 1980). This will be achieved by increasing the range of lake water level fluctuation ($\pm$ 1.5m). The project also includes, among other things, the construction of a continuous embankment around the lake for flood control, measures to cope with the difficulties in fishery and navigation that may arise at low water levels, and the establishment of an extensive sewerage system in the lake's catchment area to prevent further degradation of water quality.

A national law, specially enacted for the implementation of this project, is the first one in Japan in which the necessity of maintaining water quality in lakes is explicitly stated. Conservationists criticize it severely because of excessive destruction of littoral ecosystems due to bank construction, some unreasonable aspects in the design of the sewerage system and other factors. Much controversy concerning the merits of this big project exists.

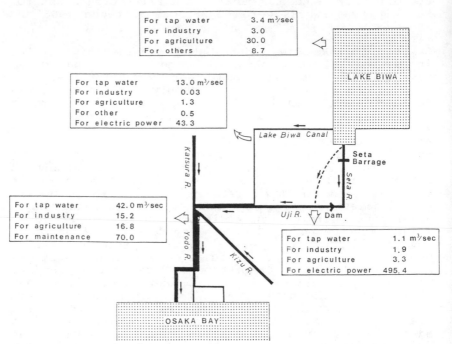

Figure 9. Current water use rates from Lake Biwa and
the Yodo River system.

WATER POLLUTION PROBLEMS

Besides supplying water to downstream industries, the
lake with its excellent water quality attracted certain
factories to establish on its shores, especially
textile industries.   Between 1917 and 1926, silk-
reeling factories became concentrated on the eastern
end of the lake.   Concurrently, one of Japan's rayon
production centers was formed in Otsu near the head of
the Seta River in 1919-1928.   In 1928, the large
quantities of wastewater discharged from the rayon
factories caused serious damage on the catch of
corbicula (Corbicula sandai).   This is a small clam
species endemic to Lake Biwa and one of the lake's main
fishery products in those days.   In response to the
fishermen's protest, the Prefectural Government of
Shiga ordered the factories to improve their sewage
disposal facilities and let the companies concerned pay
compensation to the fishermen.   This was the first
recorded event of water pollution in Lake Biwa.

Sixty years' record of transparency at the center of
the deep main basin (Northern Lake) of Lake Biwa
(Figure 10) shows a gradual decline from 1920 to 1942.
This reflects the industrial development and population
growth in the catchment area. A greater part of the
lake remained oligotrophic until the beginning of the
postwar, high economic growth period. Since the
process of rapid eutrophication that followed, has
already been reviewed in the author's address at the
Shiga Conference '84, only an example of the time trend
of organic matter increase in the water of Lake Biwa
Canal will be presented. This is illustrated in
(Figure 11). During the ten year period, from 1964 to
1973, the most drastic change in the lake water quality
took place. This was concurrent with the peaks of
industrialization in the Shiga Prefecture and Japan's
economic growth. It was also during this period that
the lake experienced the worst case of toxic
contamination with PCB, on which Mr. Fukada will report
in the Sources and Fate Session of this conference.

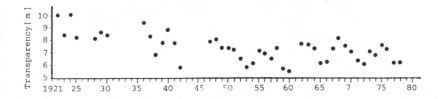

Figure 10.   Trend of transparency at the centre of the
              main basin of Lake Biwa. Drawn from the
              data   by   Shiga   Prefectural   Fishery
              Experiment Station.

The enforcement of a series of national laws on the
prevention of environmental hazards, during 1967-1970,
moderated the progress of water quality degradation to
a considerable extent in Lake Biwa as well as in other
water bodies. This is evident by the limiting trend of
the curve in Figure 11. Eutrophication of the lake
continued to advance steadily through the 1970's. This
was due to increases in wastewater input, particularly
from domestic sources. Immigration of urban population
into Shiga from neighboring districts, the rising
standard of living, and slow progress in sewerage
construction all led to increase the nutrient loading
to the lake due to domestic wastewater.

It was in response to this situation that the residents
of Shiga and their Prefectural Government took the

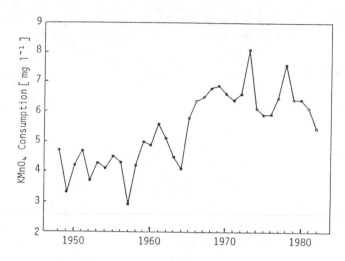

Figure 11.   Trend of organic matter concentration in
the waters of Lake Biwa canal in terms
of $KMnO_4$ consumption.   Drawn from the
data by Kyoto Municipal Waterworks
Department.

initiative in 1980 by enacting the Ordinance for
Prevention of Eutrophication in Lake Biwa.   Control of
industrial wastewater discharge was made stricter than
before and, for the first time in Japan, the use of
phosphate-containing synthetic detergents was banned.
The Ordinance proved successful.   The nutrient level in
the lake water has ceased to increase since its
enforcement.   This decision-making and experience gave
a chance for citizen leaders and government officials
in Shiga to gain a deeper insight into environmental
problems , particularly in lakes, and encouraged them
to sponsor an international conference on lake
environments so as to exchange their experiences with
those who are concerned with other lakes in the world.

I am extremely happy to be here at this conference
which succeeds Shiga's initiative, and do hope that the
spirit of the two conferences will be maintained in the
future with the expectation that these conferences will
have a significant impact on the future of our lakes.

CHAPTER 7

TRANSBOUNDARY MANAGEMENT OF LARGE LAKES: EXPERIENCE WITH THE GREAT LAKES OF NORTH AMERICA

Lee Botts

Center for Urban Affairs & Policy Research,
Northwestern University, Evanston, IL 60201

INTRODUCTION

The Great Lakes of North America divide the continent politically but unite the United States and Canada in transboundary resource management by treaty.   For three-quarters of a century, the 1909 Boundary Waters Treaty has provided a model for peaceful co-operative resource management across an international border.

Often criticized, the arrangements have functioned well enough that no suggestion for abolishing them would be seriously entertained on either side.   Limited by provisions that reflect the concerns of the times when it was developed, the treaty has yet proved far sighted and flexible enough to allow continuing evolution of resource management concepts.   The bi-national experience is the basis for growing appreciation of the need for an ecosystem approach in Great Lakes management.

For the Great Lakes, the treaty provides procedures for bi-national management of water quantity and water quality.   Concern about quantity led to negotiation of the treaty with a provision that later provided the framework for the Great Lakes Water Quality Agreement.

The Agreement is an example of imperfect arrangements for transboundary management of resources that still foster co-operation among many jurisdictions at every level of government. With the treaty and agreement as the basis for bi-national efforts, the lakes benefit from the attention of a Great Lakes community that works together across the border outside formal governmental processes.

The formal responsibility for administering the treaty and the Great Lakes Agreement rests with the International Joint Commission of Canada and the United States (IJC). This uniquely independently bi-national agency was established by the 1909 Treaty.

THE 1909 BOUNDARY WATERS TREATY

The Boundary Waters Treaty of 1909 applies to the waterways that cross the international border but is not considered to apply to the oceans. The IJC has resolved serious questions (Willhas resolved serious questions (Willoughby 1981) on rivers and other lake but has given most attention to the Great Lakes, the continent's largest reservoir of fresh water.

The Webster-Ashburton Treaty that demilitarized the Canadian-U.S. border in 1842 provided no procedures to resolve differences over competing uses of common waterways. The Boundary Waters Treaty came about because of potential disagreements over diversions that might lower water levels enough to interfere with navigation. In the last decades of the 19th century, there were also questions about withdrawals for irrigation and for production of electricity by hydropower (Dreisziger 1981).

The new Treaty was designed to set priorities for uses, to resolve conflicts, and to assist the governments in identifying new problems as they develop. Most provisions addressed water quantity issues. The first priority for use was domestic supply, including uses for sanitary purposes. Navigation was next in importance and then power production.

Though the possibility of water quality degradation was acknowledged in Article IV with language that could be broadly applied to pollution issues, pollution was then considered a local and minor matter. Nor were the drafters of the Treaty concerned with protection of either water quantity or water quality for other uses that are now considered very important. Uses not given explicit priority under the Treaty include industrial water supply, recreation, and aesthetic enjoyment.

The Treaty's designers mainly considered management of levels and flows through certain connecting channels. Still, the institutions and processes they established provided precedents that would be followed later in the Great Lakes Water Quality Agreement. Perhaps the greatest testimony to their foresight is the fact that the IJC has split along national lines in fewer than half a dozen of over 100 decisions in almost 75 years (Carroll 1983).

THE INTERNATIONAL JOINT COMMISSION

Every aspect of the International Joint Commission established by the Treaty is meant to be bi-national. Each side is expected to contribute equally to support of the IJC and has equal administrative and decision making responsibility. Locations of meetings of the IJC, or on its behalf, are alternated on each side of the border, with a representative of the host nation presiding and responsible for arrangements. This tradition is now maintained in many other governmental as well as non-governmental joint activities involving citizens of both countries.

The extraordinary structure of the IJC is designed to protect the interests of Canada, the smaller, less powerful, of the two parties. To support its participation as an equal partner, Canada contributes relatively more to the IJC in relation to its resources and population, which is only one-tenth as large as the population of the United States. In the other direction, the treaty with Canada is considered to be a remarkable example of sacrifice of sovreignty by the United States.

Each side has three IJC members, appointed by the head of the federal governments. They are charged with serving in their individual capacities rather than as representatives of the national interests. By the testimony of both observers and members, in general most IJC commissioners have accepted this charge and acted accordingly.

An IJC office is maintained in each of the two capitols with a small administrative staff and budget. Technical work is carried out and financed by government agencies whose professional staffs are also supposed to act in their personal capacities when they are doing IJC work.

Whether it is possible for individuals to separate the

interests of the agencies that employ them from the interests of the IJC is a continuing question. Yet in most cases the government experts who assist the IJC appear to operate as independently of either national or agency constraints as the IJC commissioners.

The IJC serves three principal functions, with a fourth function assigned by treaty but not performed to date. First, the IJC makes decisions for limited uses of boundary waters and approves or disapproves actions that would affect levels and flows across the border. Though this authority is often described as quasi-judicial, the function is more management and political than judicial (Munton 1979).

For the Great Lakes, protection of navigation and hydropower uses involves limited regulation of the levels of Lake Superior and Lake Ontario by control structures in the St. Marys and Niagara Rivers and in the St. Lawrence Seaway. This limited control of water quantity is the IJC's only authority to regulate directly rather than by instruction from or advice to the governments. Even so, the IJC acts directly within the terms of and to satisfy control orders adopted according to procedures specified in the treaty.

Current control orders for Lake Superior require, for example, that Lake Superior interests be balanced with those of downstream riparian property owners in decisions on the flow to be allowed through the control locks at Sault Ste. Marie. Water can be held back only as long as the level of Lake Superior does not exceed 602 feet above sea level as measured at the international datum point in Father Point, Quebec. The IJC receives technical information and advice on levels and flows from separate control boards for each channel or waterway.

Investigation at the request of the governments by formal reference is the second major function of the IJC and is linked to the third function of advice to the governments about problems or situations that need attention. In practice, the references come from both governments, although the Treaty would allow a reference to be initiated by either government alone.

When a reference is received, a bi-national study board is formed. Like the control boards, study boards consist of staff from government agencies whose areas of responsibility are pertinent to the issue to be investigated. There is a chairman for each side with equal responsibility for managing the study.

With co-ordination among so many agencies and with so

much protocol to be observed, IJC studies are cumbersome. They are often criticized because they take so long to accomplish, up to a decade in the case of the 1964 reference on Great Lakes levels and flows. Their value is also questioned because the governments are not obliged to follow or even respond to the advice received from the IJC.

Some references are a request to the IJC to advise the governments how to resolve a difference, which is not the same as the fourth, never used, function assigned to the IJC in the Treaty. To date, the governments have not referred disputes to be resolved by the IJC in binding decisions, even though this authority was given to the commission in Article X of the Treaty. The governments have often, however, taken advice requested from the IJC about how to resolve controversies over such issues as maintenance of levels or apportionment of water supplies (Willoughby 1981).

As the advisor about problems that need attention, the IJC first informed the government about water quality degradation in the Great Lakes in 1919, seven years after initiating investigation of pollution in 1912. The governments were even slower than the IJC. Fifty years later, the first Great Lakes Water Quality Agreement was signed in 1972. The Agreement expanded the IJC's resource management role in ways that are still evolving and have occupied the majority of its attention since.

DEVELOPMENT OF THE GREAT LAKES WATER QUALITY AGREEMENT

Before about 1800, the Great Lakes had changed little since European explorers found them in the 16th century. Changes in water chemistry and in the Great Lakes fishery due to intensive development and industrialization was observed in some Great Lakes locations by the end of the 19th century but their significance was not recognized. There was little concern about water quality or pollution in either country until waterborne disease became a serious problem for the rapidly growing cities.

Chicago was the first city seeking protection of its drinking water source in a Great Lake by reversing the flow of the Chicago River to carry sewage away from Lake Michigan into the Illinois and Mississippi river systems. Chicago was motivated by typhoid and cholera epidemics. Public health was also the basis for the IJC decision to investigate water quality in 1912.

Seven years later the governments ignored IJC advice that water quality degradation in the Great Lakes required attention.

By the end of the next decade, declining oxygen levels in Lake Erie were reported by scientists. Gradual increases of biochemical oxygen demand (BOD) in bays and harbors and greater algae growth received little attention in the 1930s and 1940s. In the 1950s, public concern began to grow about major fishery changes and more obvious pollution in many locations.

The International Association of the Great Lakes Researchers was organized in the late 1950s. Scientist alarmed about dramatic species changes were giving more attention to the possible causes of greater turbidity and high BOD in Lake Erie and Lake Ontario. By 1960, scientific reports said that eutrophication was being accelerated by man's activities around the western basin of Lake Erie. A news reporter's interpretation that "Lake Erie is dying" became a symbol of environmental degradation for the public in both countries.

In 1964, a reference to the IJC requested investigation of pollution in the Great Lakes and their connecting channels and advice about remedial programs. By the end of the 1960s, scientists agreed that phosphorus is the limiting nutrient for the Great Lakes and urged reduction of phosphorus inputs to slow cultural eutrophication.

When the IJC reported results of its investigation in 1970, citizens on both sides of the border were demanding governmental action against pollution generally. The Great Lakes Water Quality Agreement was negotiated at the same when the governments were adopting new national policies, laws and programs for environmental protection. The new Agreement was built on earlier experience under the Boundary Waters Treaty but called for new arrangements.

OBJECTIVES AND PROCESSES OF THE GREAT LAKES AGREEMENT

The Great Lakes Water Quality Agreement sets forth common objectives that the governments of Canada and the United States have concurred are necessary "to restore and enhance water quality in the Great Lakes System." The objectives are to be achieved by remedial programs under national laws, by separate and co-operative research and by co-ordinated surveillance and monitoring.

The Agreement provides that the objectives can be modified or altered as conditions change, as new information is obtained or as new problems develop. It is also flexible in stipulating that implementation is to be carried out both by incorporating its objectives in national programs and by co-operating in bi-national programs through the IJC.

By specifying that remedial programs will be carried out under national laws, the Agreement both respects the sovreignty and recognizes differences in the two systems of governments. In both countries, only the federal governments can make international commitments. Because the province has the primary responsibility for environmental protection in Canada, the Canada-Ontario Agreement spells out how the province will help meet national obligations with special federal funding.

In the United States, the pattern of environmental law sets minimum national standards to be achieved through state-federal partnerships. The federal government can delegate authority for implementation of environmental standards to states that meet required conditions. To date, the Environmental Protection Agency (EPA) has negotiated individually with the eight states in the drainage basin for participation in meeting obligations imposed by the Great Lakes Agreement.

The United States Constitution prohibits state participation in foreign affairs, including separate treaties and international agreements, but recent decreased federal funding for environmental programs has increased state demands for a role in setting water quality objectives under the Agreement. The organization of the Council of Great Lakes Governors with participation by the premiers of Ontario and Quebec reflects a trend toward direct participation in bi-national activities across the border outside formal arrangements between the two federal governments.

The basic approach to pollution control is also different in each country. In Canada, the provincial government has substantial power to negotiate privately with the producers of pollution for both the means and the levels of control to achieve ambient conditions. In the Clean Water Act of the United States, the minimum national standards are technology-based and there are requirements for opportunity for public participation in both standard setting and enforcement.

Both countries established new Great Lakes research programs to meet the research requirements of the Agreement. This also stimulated university and private research activities. In the international scientific community it is difficult to distinguish research initiated solely to satisfy requirements to the Agreement because so much interest in the Great Lakes has been generated by the information exchange that the Agreement fosters.

Most Canadian government research is carried out through the Canada Centre for Inland Waters by both provincial and federal agencies. In the United States, the EPA has the lead responsibility for implementation of the Agreement but substantial research is also carried out by the National Oceanic and Atmospheric Administration and the Army Corps of Engineers. EPA's Great Lakes National Program Office was established to promote incorporation of Agreement objectives into national remedial programs and research for water pollution control. Most research by individual states is co-ordinated with federal research, as are the monitoring and surveillance activities required by the Agreement.

The first major co-operative research program was carried out by the Pollution from Land Activities Reference Group from 1973 to 1976. The reference calling for the research was submitted to the IJC when the Agreement was signed. It posed three questions:

1. How much pollution of the Great Lakes is caused by land runoff?

2. Where it is occurring?, and

3. What should be done about it?

The governments have never formally responded to the IJC's report, but results of the study have influenced remedial programs and provided the basis for modification of the objectives in the second Great Lakes Agreement that was signed in 1978.

Surveillance and monitoring for the Agreement is carried out separately by national agencies under a co-operative plan. The purpose is both to assess progress toward meeting the objectives and to detect new and emerging problems. Like the water quality objectives and remedial programs, the approach to surveillance has evolved with better understanding of the complexities of Great Lakes problems.

Initially the emphasis was on measuring water chemistry. The surveillance schedule provided for intensive measurement of the levels of many substances in each lake by turn, over a multi-year cycle (Great Lakes National Program Office 1985). Today, the emphasis is on understanding the sources and fate of toxic contaminants and how they affect the ecosystem.

The Great Lakes Regional Office of the IJC participates more directly in co-ordination of surveillance and monitoring than in remedial programs or research. The regional office is the chief IJC agent in the implementation of the Agreement.

THE IJC AND THE GREAT LAKES WATER QUALITY AGREEMENT

The 1972 Agreement gave the IJC responsibility for overseeing implementation and called for a new regional IJC office that would have no other responsibilities except Great Lakes water quality. The office was established in Windsor, Ontario, across the river from Detroit, with its own bi-national staff of administrators and technical experts and separate funding.

The independence of the regional office became an issue when the first Agreement was revised in 1978 and its relationship with the IJC offices in the two capitals remains ambiguous (Carroll 1983, foreword). The question was whether the office should be a watchdog for the IJC and make judgments about the performance of the parties to the Agreement. It has evolved that not only does the regional office provide public relations services to the IJC, but more so provides secretariat services to the two new boards of experts that were established for the Great Lakes Agreement. The regional office exercises no direct oversight authority for how well Agreement obligations are being met except through the two boards whose annual reports are a major part of the Great Lakes Agreement process.

THE WATER QUALITY AND SCIENCE ADVISORY BOARDS

Like other IJC boards, the two boards of experts called for in the Great Lakes Agreement are bi-national. As with the levels control boards, members of the Water Quality Board represent the agencies on both sides that have the chief responsibility for the issues at hand, in this case, water pollution control. Members of the Science Advisory Board include both staffs of government research agencies and academic experts. Both boards work through numerous committees, subcommittees and task forces whose members may include members of the public and representatives of private organizations.

Both boards submit annual reports to the IJC. The board reports are supplemented by reports on specific topics and from the many special subunits. They are intended to provide the IJC commissioners with the basis for the IJC's own reports to the governments on progress toward meeting the agreement's objectives. Both the board reports and the IJC reports serve several purposes.

First, they provide a historical record of activities under the Agreement. The National Academy of Sciences and the Royal Society evaluation of results of the Great Lakes Agreement in 1985 was based largely on review of the annual reports of the two special boards and on the IJC's reports to the governments (US NRC and RS of Canada 1985).

Second, the annual reports, particularly from the Science Advisory Board and its subsidiary units, provide a forum for technical and scientific experts to raise policy issues for consideration. The issues are raised not only to the IJC itself but, because of the third function the reports serve, to a much larger audience.

The **third** function is to make the boards and indirectly the governments accountable to the public. They do this by informing politicians and the general public about Great Lakes problems under the guise of reporting progress under the Agreement.

Each year the meeting in which the board reports are presented is a gathering of members of the bi-national community that has developed around the agreement processes. Participants include individuals with special interests in Great Lakes matters and members of private citizen and environmental organizations as well as participants in the remedial, research and monitoring programs for the Agreement. At times, the IJC regional office organizes special activities to attract even more attention.

In effect, the annual Water Quality Board report is regarded as a report card to the public at large on the status of the Great Lakes. This function is enhanced by identification of specific local areas where objectives of the Agreement have not been met. Initially called "problem areas," these geographic locations are now called "areas of concern". Citizens and news reporters who are unconcerned about the abstract requirements of the international Agreement pay more attention when problems are identified in specific geographic locations with which they are familiar.

The reporting and accountability serve a fourth function, support for the international Great Lakes community that has grown larger and more active since the Agreement was signed. The community includes scientists, policy experts, agency staffs, environmental and other public interest organizations and elected officials.

Increasingly, members of this community meet and exchange information or work directly together across the border outside the formal agreement processes that initially brought them together. This community provides the nucleus for general public support for the Agreement as a means for Great Lakes cleanup, even though there is little general understanding of how the objectives of the Agreement or the processes by which they are being sought have evolved.

EVOLUTION OF IMPLEMENTATION OF THE GREAT LAKES AGREEMENT

The 1972 Great Lakes Agreement emphasized reduction of phosphorus loadings with a specific objective of a 1 mg/L effluent limit for all direct discharges. Substantial control of phosphorus discharges is the greatest success under the Agreement to date and is believed to be the reason for apparent slowing of eutrophication in Lakes Erie and Ontario and southern Lake Michigan, including Green Bay.

Initially remedial programs focused on expansion of sewage treatment capacity and reduction of the phosphate content of detergents in most Great Lakes jurisdictions. Later, results of the studies by the Pollution from Land Use Activities Reference Group led to demonstration projects for conservation tillage to reduce land runoff from agricultural lands and to a new approach to phosphorus control in the second Agreement.

The 1972 Agreement called for review of the objectives after five years. The review was carried out in 1976 and 1977 and a second Agreement was signed in 1978.

The new Agreement added target loadings to effluent limits as an objective for phosphorus. Based on the concept of mass balance, the target loading for each lake was to take into account all sources in determining the levels of control necessary to slow eutrophication. Less progress has been made toward meeting the two new objectives that were added in the 1978 Agreement and are the chief focus of Agreement activities today.

The new objectives were a call for an ecosystem approach to management and virtual zero discharge of toxic substances. The Great Lakes ecosystem is defined as **"the interacting components of air, land, water and living organisms, including man,"** within the Great Lakes watershed. Toxic substances are broadly defined as any substance that affects the health and well being of living organisms directly or by concentration in the food chain or in combination with other substances.

The new objectives were based on growing concern about toxic contamination in the Great Lakes food chain and recognition that the contaminants reach the lakes from many sources by many pathways, including the atmosphere. They also acknowledge that the size and the relatively closed nature of the Great Lakes system makes the lakes especially vulnerable to persistent organic chemicals and heavy metals that bioaccumulate in the food chain.

The 1978 Agreement called for a second review of Agreement objectives after the IJC submits the last of three biennial reports to the governments on progress toward achievement of its objectives. In 1986, as the time for the third report approaches, the lakes and many of their tributaries appear to be visibly cleaner and less polluted than when the 1972 Water Quality Agreement was signed. The research and monitoring required that the Agreement continue to provide new information both about the presence and levels of contaminants and biological effects. The question is whether the Great Lakes Agreement can facilitate bi-national action against toxic contamination as effectively as it did against eutrophication.

DISCUSSION

Many proposals have been made to remedy perceived deficiencies or to strengthen the International Joint Commission. Some proposals have addressed functional issues that would not require a new treaty. Selection and appointment of Commissioners, for example, can be changed unilaterally by either government.

Proposals that would require a new treaty include: giving the IJC supra-national authority to force the governments to follow through on IJC recommendations or new responsibility for transboundary water resource planning.   Most analysts have concluded that any changes that would require a new treaty would risk loss of the considerable capacity for bi-national resource management that has been achieved under the existing arrangements.   The argument is that the present arrangements have proved flexible enough to adapt to new problems and that it would prove impossible to negotiate a new and better treaty today.

The reduction of phosphorus loadings required many different actions by numerous jurisdictions.   Though there have been failures, such as the lack even now of a phosphate detergent ban in the State of Ohio, collectively local, state, provincial and federal governments and the citizens on both sides have contributed to the slowing, perhaps even the reversal, of cultural eutrophication of the largest fresh water system on the globe.

Now the mass balance concept is being applied to the development of remedial programs for toxic contamination.   Research is now underway to increase understanding of how toxic contamination affects biota and of the sources and fate of toxic contaminants in the ecosystem. With the need for an ecosystem approach to management as the basis, new attention is being given to relationships between land, air and water.

The surveillance and monitoring program now includes a Great Lakes Atmospheric Deposition (GLAD) network of stations to measure toxic substances in the air.   The data are needed to develop controls for atmospheric deposition of toxic substances into the Great Lakes but will be integrated with acid rain transport data to provide to much more comprehensive information about North American air quality.

The willingness and capacity of the bi-national Great Lakes community to act outside the formal arrangements of the Treaty and the Great Lakes Agreement continues to grow.   The governors and the premiers demonstrated their common concerns by signing a Great Lakes Charter in 1985.   In this first regional Agreement that was developed without the participation of the federal governments, they agreed to consult with each other on diversion and water quantity issues.   Now they are reportedly ready to demonstrate their common concern by signing a statement of principles on management of toxic contamination.

The State of New York, the Province of Ontario, Environment Canada and the Environmental Protection Agency joined in a separate Agreement to study toxic contamination of the Niagara River. These agencies and several others now are participating in a comprehensive study of pollution problems in the connecting channels to assist development of remedial action plans in the areas of concern that have been identified by the Water Quality Board. The connecting channels study thus serves but is not part of the agreement process managed by the IJC.

The science establishments of both countries jointly reviewed progress under the Great Lakes Agreement through the Royal Society of Canada and the National Academy of Sciences of the United States. Regional agreement on management principles was reached through the bi-national Great Lakes Fishery Commission.

Three times, in 1972, 1976 and 1977 and again from 1982 to 1984, university faculty members from both sides have organized a series of regional seminars to consider and made policy proposals for resource management issues. One bi-national private organization, Great Lakes Tomorrow, developed a public education program called Great Lakes Decisions that has been financed and carried out on both sides of Lake Erie.

Great Lakes United is a bi-national citizens organization that promotes joint lobbying of governments by citizens across the border. Bi-national research on the possibilities for legal action across the border by citizens has been sponsored by the Canadian Environmental Law Research Foundation. The Center for the Great Lakes carries out private policy research from its offices in Chicago and Toronto. Neither time nor space permits a comprehensive listing of examples of the trend toward bi-national action outside the foreign policy institutions of government but they are proliferating.

CONCLUSION

Transboundary management of the Great Lakes of North America depends on both governmental and non-governmental participation in many forms. Experience under the Boundary Waters Treaty of 1909 and the Great Lakes Water Quality Agreements of 1972 and 1978 have fostered a bi-national community that participates in resource management inside and outside formal arrangements between the governments. Shared success in reducing phosphorus loadings may thus provide a basis for successful management of the more difficult problem of toxic contamination.

REFERENCES

Carroll, J. 1983. Environmental Diplomacy: An Examination and a Perspective of Canadian-US Transboundary Environmental Relations. University of Michigan Press, Ann Arbor.

Dreisziger, N.F. 1981. Dreams and Disappointments. In: The International Joint Commission Seventy Years. Spencer, R., Kirton, J. and K.R. Nossal (eds.). Centre for International Studies, University of Toronto.

Great Lakes National Program Office 1985. U.S. Environmental Protection Agency: Five year program strategy, 1986 - 1990. Chicago, IL.

Munton, D. 1979. The political roles of the International Joint Commission. Paper prepared for the Centre of International Studies, University of Toronto.

National Research Council of the United States and the Royal Society of Canada 1985. The Great Lakes Water Quality Agreement: An Evolving Instrument for Ecosystem Management. National Academy Press, Washington, D.C..

Willoughby, W.R. 1981. Expectations and Experience 1909-1979. In: The International Joint Commission Seventy Years. Spencer, R., Kirton, J. and K.R. Nossal (eds.). Centre for International Studies, University of Toronto.

CHAPTER 8

TRANSBOUNDARY MANAGEMENT OF LAKE GENEVA

Peter A. Spoerli

Chef de la division d'Assainissement et d'Exploitation, Departement des Traveaux Publics, Republic et Canton de Geneve, Suisse

SUMMARY

Based on the situation of lake Geneva between two countries, just at the northwestern boundary of the Alps and its other essential characteristics, the functioning of its international commission for water protection is shown.   Even though it is not easy to prove so scientifically, the author tries to show that the Commission has a positive effect on lake Geneva water quality.   In spite of a substantial increase in population and activities, it has been possible in the last years to contain the increase of pollution loads to the Lake.

INTRODUCTION

Lake Geneva, whose official name is "Lac Leman" is situated on the border between France and Switzerland. In order to co-ordinate the water protection activities of the two border countries, the most relevant executive organizations on either side have delegated representatives to the international Commission for water protection of Lake Geneva (Commission international pour la protection des eaux du lac Leman) abbreviated CIPEL.   The objective of this paper is to illustrate how this Commission operates and what results it has achieved. Before doing so, some general information on Lake Leman is appropriate.

RELEVANT GENERAL INFORMATION ON LAKE GENEVA

Lake Geneva is one of the biggest lakes in Europe. It's shown in Figure 1, it is approximately 65 km long and about 12 km wide. Its biggest depth is just under 400 meters. The water volume is estimated at 90 cubic km. Its main water source is the Rhone river which enters it from the east end. The same river is the only outlet for the Lake at Geneva at its western end. This river has its origin in the central Swiss Alps in the form of the Rhone glacier and it ends in Marseille on the French Riviera where it flows into the Mediterranean Sea.

As far as the hydrology is concerned, one important factor to mention is the average residence time of the water in the lake. It is about 11 years. The Rhone river above the lake runs through a valley which is entirely within the Swiss Alps. Practically all the tributaries originate from the different glaciers abounding in this region. Intensive agriculture is practised on the slopes and in the plane of this valley with a particular accent on wine and fruit growing. On the industrial side, agrochemicals, aluminum refining, fine chemicals and oil refining are the main activities. Finally the higher slopes are well known for their skiing and mountaineering resorts.

As shown in Figure 2, the eastern third of the Lake is still within the last ranges of the Alps, whereas the rest is in slightly hilly country between the Jura mountains and northern foothills of the French Alps. This densely settled part comprises the French city of Evian on the southern shore and the Swiss cities of Lausanne and Geneva on the northern shore and at its western end, respectively. Here light industrial and commercial activities abound coupled with intensive and varied agriculture. Probably because of its pleasant country side and its political stability this region has also been able to attract numerous international activities which have their center of gravity at the western end of the lake, in Geneva.

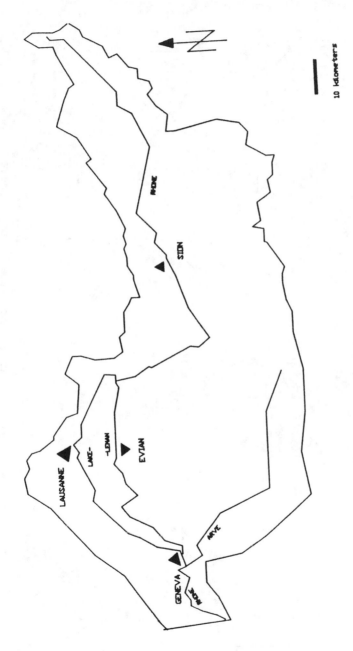

Figure 1. Lake Leman (Geneva) and its hydrological basin.

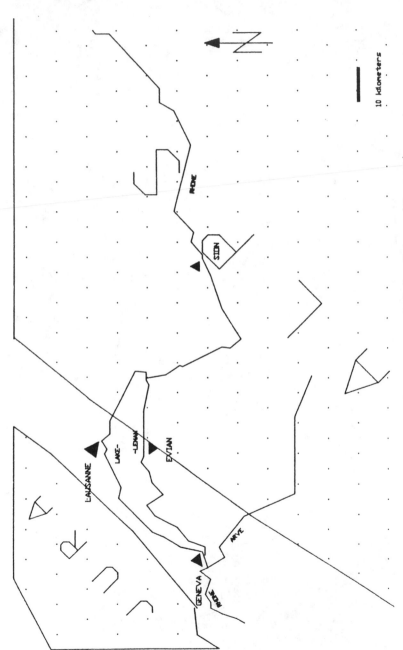

Figure 2.   Mountain ranges around Lake Leman (Geneva).

NEIGHBORS

As mentioned earlier, France and Switzerland share Lake Leman (Figure 3). To understand the functioning of the Lake Leman Water Protection Commission it is perhaps useful to mention a certain number of pertinent characteristics of these two nations.

France is a country of 50,000,000 inhabitants. Structurally it can be characterized by strong centralization which is effective on a political as well as commercial level. No city in France can rival the splendour of Paris. All important things tend to happen in Paris. Tax money flows to Paris and is redistributed from there. Politically and administratively the country is divided into somewhat less than 100 Departments. For Lake Leman the two departments of the AIN and the HAUTE-SAVOIE are relevant. Until recently financial and technical decisions were taken at the department essentially by the prefect, the representative of the national government in Paris. In recent years this centralization has, as far as financial decisions are concerned, been somewhat relaxed in favor of locally elected political bodies (General Councils).

Switzerland is quite a contrast to this. With its 6,000,000 inhabitants, it has a federal structure based on about 25 Cantons which have their local governments co-ordinated by a federal government. Needless to say that the local governments have very diversified opinions and they or the local communities collect income tax and therefore, control tax money needed to implement their decisions. Unfortunately, this situation is changing due to an ever increasing concentration of commercial decisions in a centralized location (Zurich). The Swiss cantons bordering on Lake Leman are VALAIS, VAUD and GENEVA.

LAKE LEMAN WATER PROTECTION AND THE INTERNATIONAL COMMISSION (CIPEL)

Until the beginning of this century and particularly World War I, national boundaries in this region of Europe where quite open. Frontier problems where resolved relatively easily locally. Pollution problems in the Lake Leman basin where mainly of the sanitary type, that could be resolved by the extension of sewerage systems to prevent immediate contamination of drinking water intakes. After the Second World War,

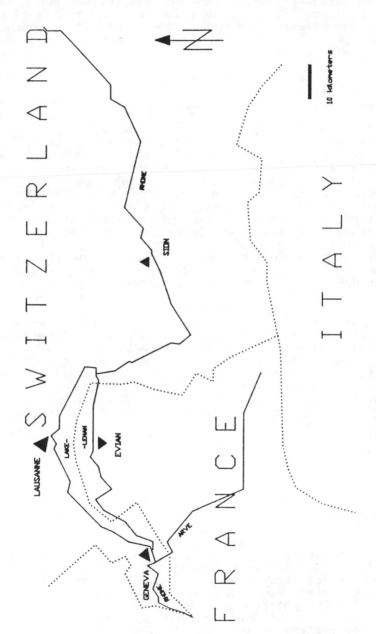

Figure 3.   Political map of Lake Leman (Geneva).

the population in the Lake Leman basin increased rapidly, concurrent with an expanding economy. The Second World War had produced national boundaries that had become quite impermeable in the meantime.

The creation of our international Commission falls into this period. In fact, it is a private organization, the General Union of Rhodanians. Its' objective was the regrouping of the inhabitants of the entire Rhone valley. This started in 1950. At that time it was suggested that a commission be created, to study the effects of sewage of the waters of the Rhone valley. This private Commission started work in 1952. Its' members where scientists of different fields most of whom worked for public administrations in France or Switzerland. In 1957 this Commission started a systematic observation of water quality of the Lake. These observations have continued ever since. Very quickly the unofficial status of the Commission turned out to be a disadvantage. Therefore, in 1960 negotiations on a diplomatic level where started in order to give the Commission official status. Two years later an Agreement was signed which after ratification was implemented at the end of 1963.

ESSENTIAL TASKS OF THE COMMISSION (CIPEL)

Four main tasks mentioned in the Agreement are:

1. To organize and have all research carried out necessary to the understanding of the nature, the importance and the sources of pollutions and to exploit the results of this research.

2. To recommend to the contracting governments what measures to take to curb existing pollution and prevent all future pollution.

3. The commission can prepare the elements of an international regulation on the subject of water quality in Lake Leman.

4. It examines any other issue pertinent to water pollution.

ORGANIZATION AND COMPOSITION OF THE COMMISSION (CIPEL)

One of the basic principles of international commissions on any subject is of course that the member countries are treated on an absolutely equal basis. That is the reason that the presidency of the Leman Commission is alternately Swiss and French. For the period of 1985 to 1987 the president is a member of the French ministry of foreign relations. Whereas for the Swiss periods it is usually a high official of the federal environmental protection office. This difference in functions reflects some of the structural differences between the countries.

This difference is again visible in the delegations. Whereas the French delegation is practically only composed of local administrative and technical representatives of the national government, the Swiss delegation is composed of only three delegates of the federal governments. The six remaining delegates are elected executive members of the governments of the three coastal cantons. Therefore, one finds on one side high civil servants and on the other a majority of professional politicians holding executive office.

For its own administrative work the Commission has a permanent secretariat established in Lausanne, with a full-time staff.

A number of working groups comprising specialists who are not members of the Commission are intended to do or follow most of the scientific or technical work.

Amongst these there is a working group for mutual assistance in case of oil spills whose task is the adaptation of the international agreements on this subject, and the implementation of these on the organizational levels. Transmission of alarms, conditions of border passage for the pollution fighting equipment and men and standardization of equipment have been their main tasks. The members of this group are mainly local fire chiefs. Oil spill clean-up is one of their tasks.

The biggest working group is the "Technical Sub-Commission". It is essentially the continuation of the original private Rhodanian commission. Four working groups are active in the subcommission whose principal activities are:

1. Defining methodology for analytical work

2. Isolation and control of specific sources of pollution

3. Estimating effects of diffuse pollution

4. Defining a five year plan and issuing of commission reports

This is the main frame of the organization. In this paper two further working groups have been intentionally omitted. One which deals with the Commission's finances and another that distributes subsidies for phosphate elimination.

Basically the Commission has an upper governing body whose members are high government officials on the French side, and a majority of elected executive politicians on the Swiss side. The actual current technical and scientific work is in the hands of the technical subcommission whose members are exclusively specialists employed by the administrations of the various border states.

FINANCIAL ASPECTS

The Commission works on a budget paid by the member states. For 1986, 802,000 - sFR. have been provided. Of this an estimated 255,000 - sFr. have been allocated to internal cost such as the permanent secretariat, commission sessions and publishing reports. The rest, namely 547,000 - sFr. are to be spent for financing commissioned work.

THE WORK OF THE COMMISSION

From here on this paper will be less factual. If I came all the way from Europe to your congress, it was to bring you more than what was officially published. In fact, what I will give you here is a personal appreciation of the achievements of the Commission. The appreciation is personal because of the difficulty of directly measuring this achievement. The water quality of Lake Leman for instance has been influenced by the Commissions' work, but at the same time demographic and economic development have taken place so that it is very difficult to distinguish the effect of each one of these influencing factors.

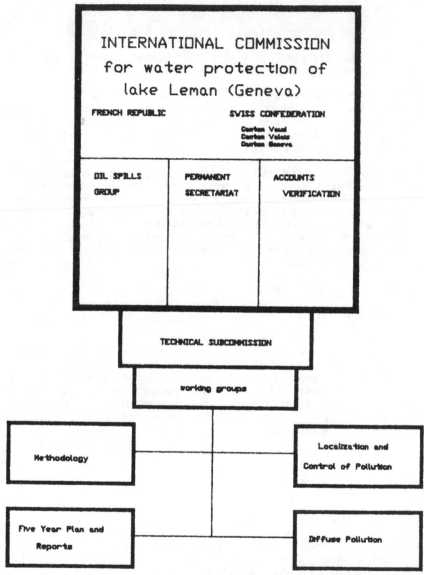

Figure 4.   Organization of the Commission.

But before sharing with you my impressions, I will try to give you some factual information on how the Commission works and what it has done.

The International Commission has no direct power of decision. The only thing it can do is recommend to its member states what action should be taken. It is up to these states to implement these recommendations within their jurisdiction. This may seem to be little or nothing. However, the fact that practically all locally responsible organizations are represented with important people in the Commission, ensures that the recommendations are taken seriously. That is of course also the reason why the recommendations never go beyond what each delegate of the governing body can accept. So the recommendations are feasible. Finally they are published. These publications have a very immediate effect on ecologists, who are an important political force at least on the Swiss side of the border. Political pressure is then exerted on the local governments to make a maximum effort to implement the Commissions' recommendations. For example the 1985 recommendations were:

1. to extend the agreement on phosphate reduction, because of its positive effects,

2. to generally promote the use of detergents without phosphates,

3. to promote the construction of separate sewage collection systems and to repair and upgrade existing sewerage systems,

4. to reinforce quality control of effluents of sewage treatment plants by providing the responsible organizations with sufficient resources, and

5. to undertake all efforts to reduce diffuse pollution by encouraging rational use of fertilizer and manure and by promoting ways of cultivation that will reduce soil erosion.

The application of these recommendations will take some time, depending how motivated the different states are. Obviously, people living far upstream of the Lake are less motivated that the fortunate ones who live on its shores.

In addition to the actual recommendations, the commission publishes annual reports about chemical and biological water and sediment quality as well as the results of quality control of sewage treatment plants.

The 1984 report also gives the first results of measurements of diffuse pollution. In 1985 a five year report was published for the first time giving detailed information on the evolution of Lake Leman water quality. These publications have quite an impact on the political level, at least as long as they show a deterioration of water quality.

As promised earlier I will try to give some indication on what I consider as being the effects of the Commission's work. For those who have followed the commissions work its publicized battle against heavy metals and especially mercury come immediately to mind (Figure 5). Thanks largely to the Commission publications, the contribution from the very few industrial point sources have been drastically reduced.

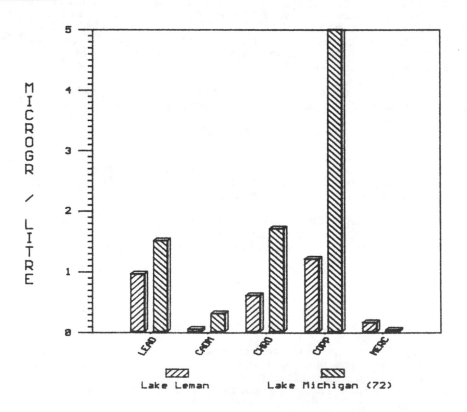

Figure 5.   Lake water metal concentrations.

The fact that 80% of the population of the Lake Leman basin is connected to sewage treatment plants and that 70% have their wastewater treated to also eliminate phosphates is not entirely the Commissions' achievement.   However, without the Commission the percentages would be lower, and especially the quality of treatment would generally be lower.   Today we begin to see signs of diminishing phosphorus content of Lake Geneva (Figure 6).   If this tendency is confirmed, the achievement can certainly be credited largely to the International Commission.

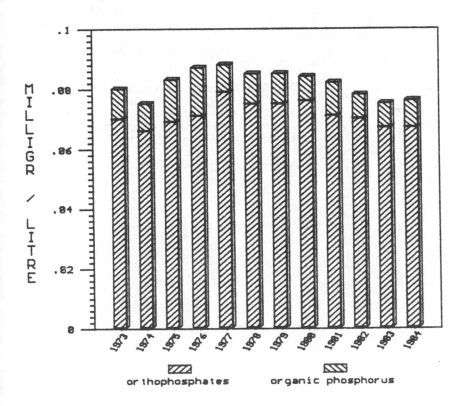

Figure 6.   Phosphorus evolution.

In Switzerland, phosphate contents in household detergents have been progressively reduced until they were completely banned for textile detergents.   This is a result of much publicized political discussions on the basis of the CIPEL reports.

A few years ago an agreement was signed between the canton of Geneva and other states providing a subsidy to be paid to sewage treatment plant operators on the basis of the attained effluent quality. This leads to regular measurements of effluent quality and the publication of the results. All this should tend to improve water quality.

It is in the forum of CIPEL that another idea has grown, that of diffuse pollution. Since the main source of this type of pollution is agricultural activities and farmers on the other hand are important politically, it has taken quite a few years of argument and persuasion to get to the point where everybody agrees that steps should be taken to reduce the negative effects of our very high intensity agricultural activity.

As a closing remark I must mention that the International Commission for Lake Leman has been largely criticized for inactivism. However, I believe that this has not been quite fair. I hope that the few examples of its activity and its effects that I have cited will make you agree with me.

CHAPTER 9

MODEL-BASED DECISION SUPPORT SYSTEMS: APPLICATION TO
LARGE LAKES AND HAZARDOUS WASTE MANAGEMENT

Kurt Fedra

International Institute for Applied Systems Analysis
(IIASA), A-2361 Laxenburg, Austria

ABSTRACT

Large lakes are recipient environmental systems for the
waste streams from numerous interdependent human
activities. Integrating large and complex physical and
at the same time socio-economic catchment areas, their
environmental problems are an intricate mixture of
physical, biological, technological, economic, and
ultimately political causes and their relationships.
The vast amount of complex and largely technical
information and the confounding multitude of possible
consequences and actions taken on the one hand, and the
complexity of the available scientific methodology for
dealing with these problems on the other, pose major
obstacles to the effective use of technical information
and scientific methodology by decision makers.

This paper describes the aims, scope, and design
philosophy of a new generation of model-based decision
support systems for large socio-technical and
environmental systems. These interactive, hybrid
systems combine data base management, system
simulation, operations research techniques such as
optimization, interactive data analysis, elements of
decision technology and artificial intelligence, with a
menu-driven symbolic and display-oriented user
interface. The approach combines quantitative
numerical methods with qualitative heuristic

descriptions and is designed to give the user direct and interactive control over the system's function. Human knowledge, experience and judgement are integrated with formal approaches into a tightly coupled man-machine system through an intelligent and self-explanatory user interface. Designed on dedicated microcomputer workstations, prototype implementations dealing with problems of lakes and hazardous substances are briefly described.

INTRODUCTION

Large lakes are peculiar environmental systems. On the one hand, they usually have large catchments, and since the lakes as well as their major tributaries are most often important waterways, their shorelines and areas along the tributaries tend to be densely populated and well developed, which, unfortunately, is almost a synonym for pollution. On the other hand, lakes have neither the large volumes for dilution that oceans have, nor the rapid through-flow of river reaches.

Large lakes are often divided by state or international boundaries, resulting in a segregation of causes, effects and responsibilities under independent political and administrative units.

Finally, it is worthwhile noting that, for example, of the estimated 314 million tons of hazardous wastes generated in the U.S.A. in 1983, 311 million tons were in the form of wastewater (CMA 1983).

Given the above, it is not at all surprising that there are problems with large lakes. The Great Lakes provide probably one of the best studied examples, as several hundred references to the scientific literature found in a quick and ad-hoc computer search with the combined keywords "Great Lakes" and "hazardous" or "toxic" has demonstrated (e.g. Sonzogni and Swain 1982, Eadie et al. 1983, MDC 1982, GLBC 1980, GLWQB 1978, 1981). To quote from just one example:

Lee (1980) in his report on hazardous chemicals in Lake Ontario, based on the data collection program of the IFYGL, concludes:

> 'Summaries indicate that almost all reported measurements for total cadmium exceed the IJC water quality objective by at least five times. Values reported for copper also show a high percentage of measurements equal to or

in excess of the recommended IJC water quality objectives. Other metals showed more regional patterns of values equaling or exceeding water quality objectives. The concentrations of PCBs, dieldrin and DDT and its metabolites in Lake Ontario water and fish were found to exceed the IJC water quality objectives and the USEPA water quality criteria.'

So we know that there is a problem and we have some vague understanding as to what at least some of its causes are. What next? One obvious response is to organize a conference.

A slightly more general, or less facetious, answer is to try to bring to bear scientific methods on the ultimately political planning and decision-making process by:

-providing background information, or the intelligence activity (in the military sense) of the decision-making process as seen by Simon (1960).

-generating alternatives and analyzing the likely consequences of feasible actions, i.e. the design activity, and

-structuring the decision-making process, or the choice activity.

In summary, there is a clear need to organize and communicate information and knowledge to facilitate policy and decision making. Clearly, computers and modern information technology will have a major role to play here.

A SYSTEMS VIEW

The human environment is obviously complex, and so is environmental problem solving since it must also consider socio-political, economic and technological elements, numerous actors, differing goals and perceptions, and more or less pluralistic value systems. As a consequence, scientific evidence, and modelling in particular, has become even larger and more complex, difficult to handle, understand, and believe in. However, the essence of modelling is simplification and re-scaling to dimensions and a degree of complexity that are easy to manipulate and are directly intelligible. If scientific evidence is to support public policy making, it not only has to be useful, it must be usable.

However, for most practical problems, the isolation of a single subsystem out of any larger functional context is extremely difficult. One can not - from a policy or management point of view - reasonably separate a lake from its watershed or use, or the physical system from its political environment, working on politically or economically unfeasible solutions, or just pushing problems around so that they become somebody else's responsibility.

Science undoubtedly has made great progress in understanding many phenomena. A very large amount of observational and experimental data have been accumulated and countless models, which can be understood as composite hypothesis describing the relationships of these data (Fedra 1983) and have been formulated, tested and applied more or less successfully to further analytical studies. Most of these activities, however, are concentrated in the domains of physics, chemistry and biology. Only rarely have there been attempts to integrate what is known to date and make that directly available in an immediately usable form as input to public policy-making and environmental management.

To a degree, the reasons are obvious: at the policy level, numerous components and their relationships have to be considered, and the scientific input is only one of several components; there is no single overall expert, no single discipline with an established set of rules. Perceptions, judgements and values, revealed and hidden, play an increasingly important role.

Any more structured approach also requires arbitrary choice based on human judgement that includes subjective values and experience. This judgement and choice, however, can be supported by structuring the available information and providing access to a set of appropriate tools, while directly involving the human decision maker(s) in the process. Computer technology is designed to facilitate the manipulation of extremely large amounts of static as well as dynamic information. Computer technology can be used to design an intelligent information and decision support system that can help bridge the gap between science and public policy.

MODEL-BASED DECISION SUPPORT SYSTEMS

The central objective of the approach described here is to design, develop and implement integrated sets of software tools, building on existing models and computer-assisted procedures. Recognizing the increasing importance of a participatory style of decision making and the need to justify public policy, these sets of tools are designed for a broad group of users. They include the non-technical users. The primary purpose is to provide easy access and allow efficient use of methods of analysis and information management which are normally restricted to a small group of technical experts. The use of advanced information and data processing technology allows a more comprehensive and interdisciplinary view of the management of hazardous substances and environmental target systems. Easy access and use, based on modern computer technology, software engineering, and concepts of Artificial Intelligence (AI) now permit a substantial increase in the group of potential users of advanced systems analysis methodology and thus provide a powerful tool in the hands of planners, managers, policy and decision makers and their technical staff.

To facilitate the access to complex computer models for the casual user, and for more experimental and explorative use, it also appears necessary to build much of the accumulated knowledge of the subject areas into the user interface for the models. Thus, the interface has to incorporate knowledge-based or experiment systems that are capable of assisting any non-expert user to select, set up, run and interpret specialized software. By providing a coherent user interface, the interactions between different models, their data bases, and auxiliary software for display and analysis become transparent for the user, and a more experimental, educational style of computer use can be supported. This greatly facilitates the design and evaluation of alternative policies and strategies for the management of environmental pollution.

Conceptually, the main elements of the system are:

-an **Intelligent User Interface**, which provides easy access to the system. This interface must be attractive, easy to understand and use, error-correcting and self-teaching, and provide the translation between natural language and a human style of thinking to the machine level and back. This interface must also provide a largely menu-driven conversational guide to the system's usage, and a number of display and report generation styles, including color graphics and linguistic interpretation of numerical data;

-an **Information System**, which includes the system's Knowledge and Data Bases as well as the Inference Machines and Data Base Management Systems, which not only summarize application and implementation-specific information, but also contain the most important and useful domain-specific information, They also provide the information necessary to infer the required input data to run the models of the system and interpret their output. The Inference and Data Base Management Systems allow a context and application-oriented use of the knowledge base.   These systems should not only enable a wide range of questions to be answered and find the inputs and parameters necessary for the models, but must also be able to explain how certain conclusions were arrived at;

-the **Simulation System** consists of a set of models (simulation, optimization), which describe individual processes that are elements of a problem situation, perform risk and sensitivity analyses on the relationship between control and management options and criteria for evaluation, or optimize plans and policies in terms of their control variables, given information about the user's goals and preferences, according to some specified model of the system's workings and rules for evaluation.

-the **Decision Support System**, which integrated at various levels of the system, assists in evaluating ranking and choosing alternatives described by a large set of mostly incommensurate criteria.

These elements are transparently linked and integrated. Access to this system of models is through a conversational, menu-oriented user interface, which employs natural language and symbolic, graphical formats as much as possible. The systems must provide not only a low-cost entry for the casual user, but also have the potential to be custom configured for day to day use by users of growing expertise. Detailed technical descriptions are given in Fedra (1985, 1986a).

EXPERT SYSTEMS AND DECISION SUPPORT

Underlying the concept of decision support systems in general, and expert systems in particular, is the recognition that there is a class of (decision) problem

situations that at the outset are not well understood
by the group of people involved. Such problems cannot
be properly solved by a single systems analysis effort
or a high structured computerized decision aid. They
are neither unique - so that a one-shot effort would be
justified given the problem is big enough - nor do they
recur frequently enough in sufficient similarity to
subject them to rigid mathematical treatment. They are
somewhere in between. Due to the mixture of
uncertainty in the scientific aspects of the problems,
and the subjective and judgemental elements in its
socio-political aspects, there is no wholly objective
way to find a best solution.

One approach to this class of under-specified problem
situations is an iterative sequence of systems analysis
and learning generated by (expert or decision support)
system use. This should help shape the problem as well
as aid in finding solutions. Key ingredients, are the
Decision Makers or Problem Owners, Decision Support
Technology (which helps to express value judgements,
and formalize time and risk preferences, and tradeoffs
among them), and Information Technology, which provides
and organizes substantive background information, data
and models (see for example, Phillips 1984, Keeney
1986).

There is no universally accepted definition of decision
support systems. Almost any computer-based system,
from data base management or information systems via
simulation models to mathematical programming or
optimization, could possibly support decisions. The
literature on information systems and decision support
systems is overwhelming, and growing (e.g. Zhao et al.
1985). Approaches range from rigidly mathematical
treatment, to applied computer sciences, management
sciences or psychology.

Decision support paradigms include predictive models,
which give unique answers but with limited accuracy or
validity. Scenario analysis relaxes the initial
assumptions by making them more conditional, but at the
same time more dubious. Normative models prescribe how
things should happen, based on some theory, and
generally involve optimization or game theory.
Alternatively, descriptive or behavioural models
supposedly describe things as they are, often with the
exploitation of statistical techniques.

Most recent assessments of the field, and in particular
those concentrating on more complex, ill-defined,
policy-oriented and strategic problem areas, tend to
agree on the importance of interactiveness and the
direct involvement of the end user. Direct involvement

of the user results in new layers of feedback structures. The information system model is based on a sequential structure of analysis and decision support. In comparison, the decision support model implies feedbacks from the applications, e.g. communication, negotiation, and bargaining onto the information system, scenario generation and strategic analysis.

Often enough however, the problem holder (e.g. a regulatory agency) is not specialized in all the component domains of the problem (e.g. industrial engineering, environmental sciences, toxicology, etc.). Expertise in the numerous domains touched upon by the problem situation is therefore as much a bottleneck as the structure of the decision problem. Building human expertise and some degree of intelligent judgement into computer software is, after all, one of the major objectives of AI. Only recently has the area of expert systems or knowledge engineering emerged as a road to successful and useful applications of AI techniques (e.g. Fedra and Otway 1986).

THE PROBLEM AREA: HAZARDOUS SUBSTANCES AND
ENVIRONMENTAL QUALITY

More than 300 million tons of hazardous waste are produced annually in the U.S. For comparison, about 2 gigatons of waste are produced annually in the countries of the EC, somewhat less than 10% of which is from industrial sources. Roughly 10% of these industrial wastes are classified as hazardous. More graphically, this amounts to 20 million metric tons, or a train of roughly 10,000 km length (Schneider 1984.)

The effective management of these wastes calls for:

-a minimization of waste production by process modification and recycling;

-appropriate treatment, i.e. the conversion to non-hazardous forms;

-finally, a safe disposal of whatever is left.

In addition to hazardous wastes, there is a large number of commercial products that are also hazardous. Their production, transportation and use - all before they enter any waste stream - is also of concern. Industrial production processes that involve hazardous raw materials, feedstocks or interim products, which may reach the environment after an accident, causing direct health risks to people are of particular concern. Accidents as Seveso, Bhopal and Chernobyl are examples. As a special category, although implied in the above, transportation of hazardous substances (including, of course, hazardous wastes) is included in the system.

The entire life-cycle of hazardous substances (Figure 1) from their production and use to their processing and disposal, involves numerous aspects and levels of planning, policy and management decisions. Technological, economic, socio-political and environmental considerations are involved at every stage of the management of these life cycles and involve various levels. They range from site or enterprise to local, regional, national and even international scales, and over different times scales, from immediate operational decisions to long-term planning and policy problems.

▣IIASA Demonstration Prototype: Hazardous Substances Risk Management 𝕗𝕚𝕝𝕖 ▣

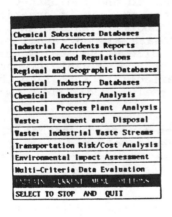
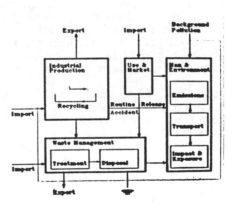

to select a menu item, position the mouse pointer,
and press the left mouse button ...

Figure 1.   Hazardous substances risk management.

The problems of managing hazardous substances are neither well defined nor reducible to a small set of relatively simple sub-problems. They always involve complex trade-offs under uncertainty, feedback structures and synergistic effects, non-linear and potentially catastrophic systems behaviour - in short, the full repertoire of a real-world mess. The complexity and ill-defined structure of most problems makes any single method or approach fall short of the expectations of potential users. The classical, mathematically oriented, but rigid methods of Operations Research and Control Engineering, that require a complete and quantitative definition of the problems from the outside are certainly insufficient.

While only the combination of a larger set of methods and approaches holds promise of effectively tackling such problems, the subjective and discretionary human element must also be given due weight. This calls for the direct and interactive involvement of users, allowing them to exert discretion and judgement wherever formal methods are insufficient (Fedra and Loucks 1985).

Under contract to the Commission of the European Communities' Joint Research Centre, (ISPRA), IIASA's Advanced Computer Applications Project is developing an interactive, computer-based decision support and information system. Recognizing the potentially enormous development effort required and the open-ended nature of such an undertaking the project is organized as a cooperative effort that takes advantage of the large volume of scientific software already available. A modular design philosophy allows us to develop individual building blocks, which are valuable products in their own right, and to interface and integrate them in a flexible framework easily modifiable with increasing experience of use.

With the functional and problem-oriented, rather than structural and methodological design of this framework, working prototypes that allow us to explore the potential of such systems can be structured at relatively low cost and with only incremental effort.

Any comprehensive assessment of the management of industrial risk, and hazardous substances in particular, requires the consideration of technological, economic, environmental and socio-political factors. Every scenario for simulation or optimization, defined interactively with this system must ultimately be assessed, evaluated and compared with alternatives in terms of a list of criteria.

These criteria include economic, technical, environmental, resource use and finally socio-political considerations.

Clearly, only a small subset of these criteria may be
expressed in monetary, or even numerical terms. Most
of them require the use of linguistic variables for a
qualitative description. Using fuzzy set theory,
qualitative verbal statements can easily be combined
with numerical indicators for a joint evaluation and
ranking. In the system design, the use of programming
languages like LISP and PROLOG gives the user freedom
to manipulate systems and numbers within a coherent
framework.

INFORMATION MANAGEMENT AND DECISION SUPPORT

The sheer complexity of the problem related to the
management of hazardous substances and related risk
assessment problems calls for the use of modern
information processing technology. However, most
problems that go beyond the immediate technical design
and operational management level involve as much
politics and psychology as science.

The software system described here is based on
information management (Figures 2,3,4) and model-based
decision support. It envisions experts, decision and
policy makers, as its users. In fact, the computer is
seen as a mediator and translator between expert and
decision maker, between science and policy. The
computer is thus not only a vehicle for analysis, but
even more importantly, a vehicle for communication,
learning and experimentation.

The two basic, though inseparably interwoven elements
are to supply factual information based on existing
data, statistics and scientific evidence and to trace
the likely consequences of new plans. The framework
foresees the selection of criteria for assessment by
the user and the assessment of scenarios or alternative
plans in terms of these criteria. The evaluation and
ranking is again done partly by the user, where the
machine only assists through the compilation and
presentation of the required information.
Alternatively, it can be done by the system on the
basis of user-supplied criteria for screening and
selection.

The selected approach for the design of this software
system is eclectic as well as pragmatic. We use proven
or promising building blocks and use available modules
where we can find them. We also exercise
methodological pluralism: any "model", whether it is a
simulation model, a computer language, or a knowledge

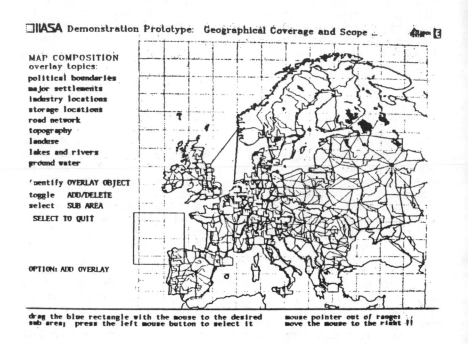

Figure 2.   Geographical coverage and scope.

of representation paradigm, is by necessity incomplete. It is only valid within a small and often very specialized domain. No single method can cope with the full spectrum of phenomena, or rather points of view, called for by interdisciplinary and truly applied science.

The direct involvement of experts and decision makers shifts the emphasis from a production-oriented "off line" system to an explanatory, learning-oriented style of use. The decision support and expert system is as much a tool for the expert as it is a testing ground for the decision maker's options and ideas.

In fact, it is the invention and definition, i.e. the design of options that is at least as important as the estimation of the their consequences and evaluation. For planning, policy and decision making, the generation of new species of ideas is as important as

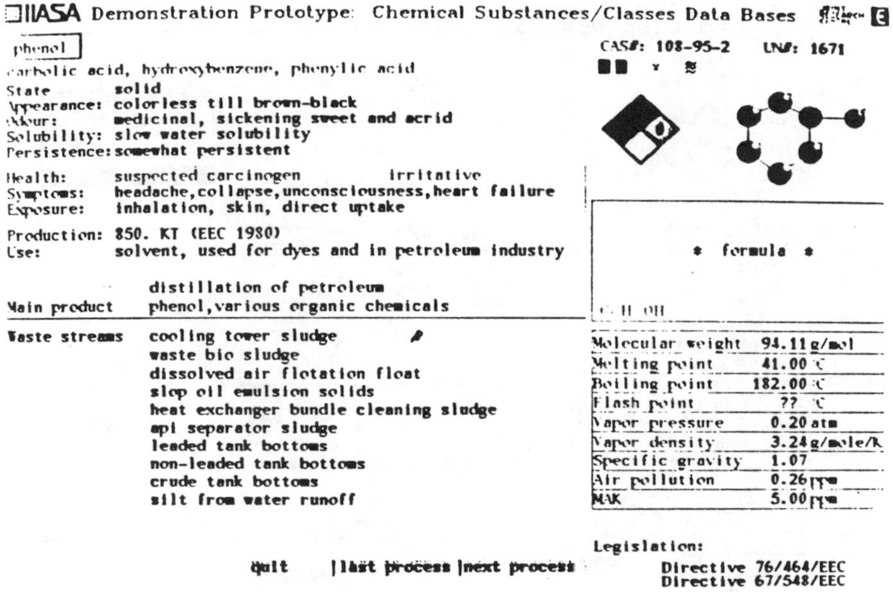

IIASA Demonstration Prototype: Chemical Substances/Classes Data Bases

CAS#: 108-95-2   UN#: 1671

phenol

carbolic acid, hydroxybenzene, phenylic acid

State   solid
Appearance: colorless till brown-black
Odur: medicinal, sickening sweet and acrid
Solubility: slow water solubility
Persistence: somewhat persistent

Health: suspected carcinogen   irritative
Symptoms: headache,collapse,unconsciousness,heart failure
Exposure: inhalation, skin, direct uptake

Production: 850. KT (EEC 1980)
Use: solvent, used for dyes and in petroleum industry

            distillation of petroleum
Main product phenol,various organic chemicals

Waste streams   cooling tower sludge
                waste bio sludge
                dissolved air flotation float
                slop oil emulsion solids
                heat exchanger bundle cleaning sludge
                api separator sludge
                leaded tank bottoms
                non-leaded tank bottoms
                crude tank bottoms
                silt from water runoff

* formula *

CLASS OIL

Molecular weight	94.11g/mol
Melting point	41.00 °C
Boiling point	182.00 °C
Flash point	77 °C
Vapor pressure	0.20 atm
Vapor density	3.24g/mole/l
Specific gravity	1.07
Air pollution	0.26 ppm
NAK	5.00 ppm

Legislation:
            Directive 76/464/EEC
            Directive 67/548/EEC

quit |last process |next process

Figure 3.   Chemical substances/classes data bases.

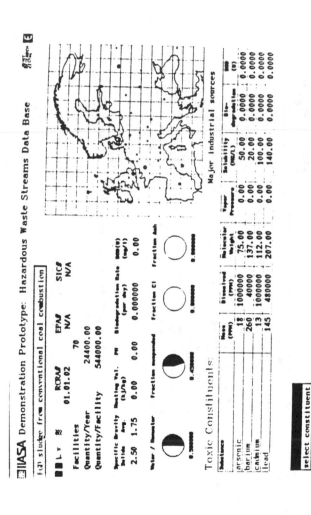

Figure 4.   Hazardous waste streams data base.

the mechanisms for their selection.   It is such an
evolutionary   understanding   of   planning   that   this
software system is designed to support.   Consequently,
the necessary flexibility and expressive power of the
software system are the central focus of development.

MODEL INTEGRATION AND THE USER INTERFACE

From a user perspective, the system must be able to
assist in its own use, i.e. explain what it can do, and
how it can be done.   The basic elements of this self-
explanatory system are the following:

   -The **interactive user interface** that handles
   the dialog between the user(s) and the
   machine; this is largely menu-driven, that
   is, at any point the user is offered several
   possible actions which he can select from the
   menu of options provided by the system.

   -a **task scheduler** or **control program**, that
   interprets the user request - and, in fact,
   helps to formulate and structure it -
   coordinates the necessary tasks (program
   executions) to be performed; this program
   contains the "knowledge" about the individual
   component   software   modules   and   their
   interdependencies;

   the control program can translate a user
   request into either:
         -a data/knowledge base query;
         -a request for "scenario analysis"

   the latter will be transferred to:

   -a **problem generator** that assists in defining
   scenarios for or optimization; its main task
   is to elicit a consistent and complete set of
   specifications from the user, by iteratively
   resorting to the data base and/or knowledge
   base to build up the **information context** or
   frame of the scenario. A scenario is defined
   by a delimitation in space and time, a set of
   (possibly recursively linked) processes, a
   set of control variables and a set of
   criteria to describe results.   It   is
   represented by

   -a set of **process-oriented models** (Figures 5-
   7), that can be used in either simulation or

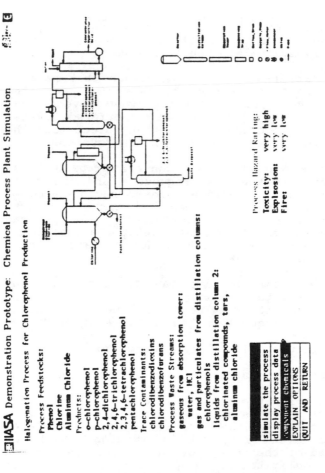

Figure 5.   Chemical process plant simulation.

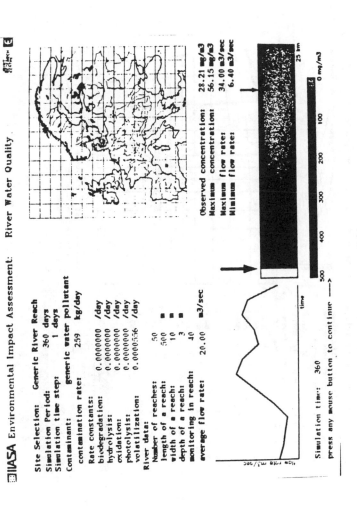

Figure 6.   River water quality.

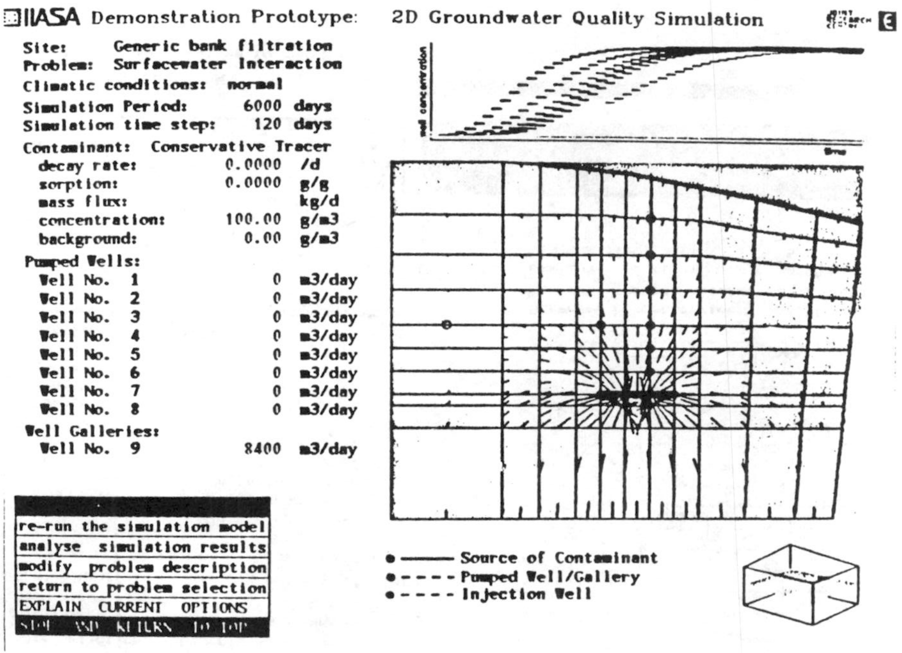

Figure 7. Ground water quality simulation.

optimization modes. The results of creating
a scenario and either simulating or
optimizing it are passed back to the problem
generator level through an

-evaluation and comparison module, that
attempts to evaluate a scenario according to
the list of criteria specified and assists in
organizing the results from several
scenarios. For this comparison and the
presentation of results, the system uses a

-graphical display and report generator,
which allows selection from a variety of
display styles and formats, and in particular
enables the results of the scenario analysis
to be viewed in graphical form. Finally the
system employs a:

-system administration module, which is
largely responsible for housekeeping and
learning: it attempts to incorporate
information gained during a particular
session into the permanent data/knowledge
bases and thus allow the system to "learn"
and improve its information background from
one session to the next.

It is important to notice that most of these elements
are linked recursively.   For example, a scenario
analysis will usually imply several data/knowledge base
queries to provide the frame and necessary parameters
transparently.   Within each functional level several
iterations are possible.   At any decision breakpoint
that the system cannot resolve from its current goal
structure, the user can specify alternative branches to
be followed.

The simulation models of the production system can be
configured to describe the comprehensive life-cycle of
hazardous substances.   The major components of the
simulation system are:

    -the industrial production sector,
    -waste management, including treatment and
     disposal,
    -the cross-cutting transportation sector,
    -man and the environment.

Each of these major components is represented by
several individual models, covering a variety of
possible approaches and levels of resolution.   Each
element of the simulation system can be used in
isolation, or is linked with several others as pre-or

post-processors into increasingly larger (sub)systems models (Fedra 1986b).

AI technology is embedded into this integrated software system at various levels, in various modules. They range from hierarchial, frame-based data bases (see Weigkricht and Winkelbauer 1986), to rule-based pre-processors and input generators for classical numerical simulation models, rule-based heuristic feasibility and consistency checking of interactive input, to symbolic simulation and intelligent parsers for language-oriented input. The emphasis, clearly, is on a broad set of problem and knowledge-representation techniques, integrated into one coherent framework.

It is important to note that none of the complexities of the system's integration are obvious to the user: irrespective of the task specified, the style of the user interface and interactions with the systems are always the same at the user end.

DATA BASES, SIMULATION AND OPTIMIZATION

The system as described can be used in a variety of ways. These modes of operation, however, serve only as design principles. They are not seen by the user, who always interacts in the same manner through the user interface with the system. The system must, however, on request "explain" where a result comes from and how it was derived, e.g. from the data- based, inferred by a rule-based production system, or as the result of a model application.

The simplest and most straightforward use of the system is as an **interactive information system**. Here the user "browses" through the data and knowledge bases or asks very specific questions (Figures 2,3,4). As an example, consider the substances data base, where the basic properties of a substance can be found (Fedra et al. 1986). The data bases can also be used from any of the impact models; the necessary substance-specific parameters are automatically retrieved and made available to the calling model.

The second mode of use is termed scenario analysis. Here the user defines a special situation or scenario (e.g. the release of a certain substance from a facility), and then traces the consequences of this situation through modelling. The system will assist the user in the formulation of these "What if..." questions, largely by offering menus of options and ensuring a complete and consistent specification.

The scenario analysis mode can use any or all models in isolation or linked together; the coupling of models is automatic.   The   evaluation   and   comparison   of alternatives is always performed in terms of a subset or all of a list of criteria, including monetary as well as symbolic, qualitative descriptors. The use of certain   models   is   implied   by   the   selection   of indicators and criteria that are chosen to describe a scenario's outcome.

Two time domains for scenario analysis with different problems addressed are supported;   the models can either be used to simulate medium- to long-term phenomena, with a characteristic time scale of years, or   short-term   events,   i.e.   accidents,   with   a characteristic time scale of days. Switching from one mode to the other, with the necessary aggregation or disaggregation of information is possible.

Similar to this switching in the time domain, a change in the space domain must also be supported. There is of course a close linkage between time and space scales, in that most short-term phenomena like spills or accidents are relevant on a local to regional scale, whereas long-term phenomena like continuous routine release   of   hazardous   substances   will   usually   be considered on a regional to national scale.

Scenario   analysis   may   be   either   straightforward simulation,   or   a   combination   of   simulation   and optimization techniques. In the latter case, the user does not have to specify concrete values for all control variables defining a scenario, but rather specifies allowable ranges on them as well as a goal structure.

Using techniques such as reference points to multi-objective   problems   (e.g.   Wierzbicki   1983),   an appropriate framework allows one to modify expectations interactively.   The user can redefine objectives and constraints   in   response   to   first   results. Alternatively, discrete optimization can be used as a post-processor for the results of simulation.   Again, interactive   selection   of   criteria,   objectives   and desired target solutions directly involves the human evaluator in the optimization process.

All of these refinements of the basic information and simulation   system   however   must   not   complicate   the users' interactions with the system.   Ease of use, and the possibility to obtain immediate, albeit crude and tentative answers to problems, the machine helps to formulate in a directly understandable, attractive and pictorial format, are seen as the most important features of the system.

AN APPLICATION TO THE GREAT LAKES?

For the Great Lakes, and within the framework of the Water Quality Agreement (NRC and RSC 1985), there are numerous problem areas where an information and decision support system such as described could be used most effectively. A high degree of complexity, a large amount of information, but also a considerable degree of uncertainty, and finally a multitude of actors involved are typical characteristics of a problem situation that can be supported by this soft and pluralistic approach to systems analysis. Many of the recommendations of the review community of the Great Lakes Water Quality Agreement could benefit from such information management. To quote just one of these recommendations:

> "...the overall objective of the Agreement should, therefore, continue to emphasize the systemic characteristics of the basin including human activities taking place in it. Four kinds of information are needed: time series of monitored data, maps of key features of the ecosystem and of its use and abuse by humans, models of causal relationships integrating human uses and ecosystems responses, and case studies of management actions to demonstrate what has worked and what has not."

Clearly all of these recommendations could benefit from a high integrated, intelligent and user-friendly support system as the above.

The open architecture and modular structure of the system described above, which should be viewed as an approach and set of tools rather than a product, allows its easy conversion and adaptation to different geographical regions and spatial resolutions.

Individual simulation modules can easily be exchanged, and in many cases the component models, based on time-tested software developed, e.g. by the USEPA, are of a general, data-driven nature that allows application to a broad range of situations by just providing the appropriate input data set.

The structure of the data bases, again is general, so that the major effort in adapting a system like the above is in compiling the appropriate data - which, in the case of the Great Lakes, at least in part, is already available in computerized form, which greatly reduces the compilation effort - and adapting part of the system's structure.

Clearly, when concentrating on a specific environmental target system, several alternative models describing this system should be integrated, and in the case of the lakes, the entire hydrological cycle should be included. Lake water quality, accumulation of toxics in the food chain, and their effects on commercial and recreational fisheries are important topics. Non-point runoff from agricultural areas, carrying pesticides, and their transport through the drainage system would have to be represented. Other components, such as the transportation risk-cost analysis module would have to be extended to handle ship transport. The specific industries in the catchment, most prominently pulp and paper, iron and steel including coking, tetraethyl lead production and general petrochemical production (NRC and RSC 1985), urban runoff, municipal treatment, effluents and advanced wastewater treatment, sludge disposal and the location and management of landfill operations in the catchment area, or atmospheric deposition of toxics, would require straightforward extensions and additions to the existing core structure and set of component models.

In all of these cases, however, numerous models and much of the require data are already available, so that the major effort is one of integration and adaptation.

In summary, the resulting system would make a large amount of background information and the full power of formal, numerical methods also available to the non-technical user, who is supposed to be knowledgeable about the problem situation and in particular its regulatory, socio-political and institutional aspects, but not necessarily about its technological components, risk analysis and computer technology. By providing easy and symbolic interaction formats, as well as directly intelligible style of output presentation, the normally very cumbersome and time-consuming task of comparative impact analysis can be greatly simplified.

Making scientifically based analysis responsive and affordable for a larger group of users, one can afford to explore many more options and alternatives than one normally would. Immediate response to questions and directly intelligible formats allow a very tight coupling of human expertise, experience and judgement, with the formal analysis and information processing of the machine. The coupled man-machine system, modelled conceptually along the lines of expert systems without making some of their excessive claims of human performance, is primarily designed to organize information. It supports decision making by providing background information, information on likely consequences of alternatives, and finally by methods that help organize, compare and evaluate complex information.

The high degree of aggregation certainly requires that much of the detailed technical information underlying the aggregated summary presentation be suppressed and only communicated to the user on specific request. On the other hand, the quick and directly understandable generation and comparative presentation of alternatives allows the decision maker to absorb the basic features of the problem situation and internalize them in terms of his own set of mental models. This we believe, is the most effective way of decision support we can offer for complex socio-technical and environmental problems.

REFERENCES

CMA 1983. The CMA hazardous waste survey for 1983. Final Report. Chemical Manufacturers Association, Washington, D.C. 35p.

Eadie, B.J., Robbins, J.A., Landrum, P.F., Rice, C.P. and Simmons, M.S. 1983. Cycling of toxic organics in the Great Lakes: A 3-Year status report. National Oceanic and Atmospheric Administration, Ann Arbor. MI NOAA/TM/ERL/GLERL-45; NOAA-83090104. 176p.

Fedra,K. 1983. Environmental modelling under uncertainty: Monte Carlo simulation. RR-83-28, International Institute for Applied Systems Analysis, Laxenburg, A-2361 Austria.

Fedra, K. 1984. Interactive water quality simulation in a regional framework: A management oriented approach to lake and watershed modelling. Ecological Modeling, 21:209-232.

Fedra, K. 1985. Advanced decision-oriented software for the management of hazardous substances. Part I: Structure and design. CP-85-18, International Institute for Applied Systems Analysis, A-2361 Laxenburg, Austria. 61p.

Fedra, K. and Loucks, D.P. 1985.   Interactive computer technology for policy modelling.   Water Resour. Res. 21:2, 143-152.

Fedra, K. 1986a.   Advanced decision-oriented software for the management of hazardous substances. Part II: A prototype   demonstration   system.   CP-86-10, International Institute for Applied Systems Analysis. A-2361 Laxenburg,Austria.

Fedra, K. 1986b.   Decision making in water resources planning: models and computer graphics.   Paper presented at the UNESCO/IHP-III Symposium on Decision Making in Water Resources Planning, Oslo, Norway, 5-7 May.

Fedra, K. and Otway, H. 1986.   Advanced decision-oriented software for the management of hazardous Substances.   Part III. Decision support and expert systems: uses and users. CP-86-14, International Institute for Applied Systems Analysis, A-2361, Austria. 44p.

Fedra, K., Mutzbauer, G., Weigkricht, E. and Winkelbauer, L. 1986.   Advanced decision-oriented software for the management of hazardous substances. Part III: Substances data and knowledge Base. CP-86-xx,   International Institute for Applied Systems Analysis, A-2361 Laxenburg, Austria (in press).

GLBC 1980a.   Great Lakes Basin plan: hazardous materials strategy.   Great Lakes Basin Commission, Ann Arbor, MI. 70p.

GLWQB 1978.   Great Lakes Water Quality Sixth Annual Report.   Appendix F.   Report on hazardous waste disposal.   Great Lakes Water Quality Board. 34p.

GLWQB 1981a.   Annual Report.   Committee on the assessment of human health effects of Great Lakes Water quality.   Great Lakes Water Quality Board. 142p.

GLWQB 1981b.   Report on Great Lakes water quality, Appendix: Great Lakes surveillance.   Great Lakes Water Quality Board. 174p.

Haile, C.L., Veith, G.D., Lee, G.F. and Boyle, W.C. 1975.   Chlorinated hydrocarbons in the Lake Ontario ecosystem   (IFYGL).   EPA-660-3-75-0021,   USEPA, Corvallis, Oregon, 28p.

Halfon, E. 1984.   Error analysis and simulation of MIREX behaviour in Lake Ontario. Ecological Modelling, 22 (1983/84), 213-353.

Keeney, R.L. 1980. Evaluating sites for energy facilities. Academic press, New York.

Keeney, R.L. and Raiffa, H. 1976. Decisions with multiple objectives: preferences and values tradeoffs. Wiley, New York.

Lee, G.F. 1980. An evaluation of hazardous chemicals in Lake Ontario during IFYGL. EPA-600/3-80-060, OSEPA, Duluth, MN., 26p.

Loucks, D.P., Kindler, J. and Fedra, K. 1985. Interactive water resources modelling and model use: an overview. Water Resour. Res., 21:2, 95-102.

Lyon, J. 1979. Remote sensing analyses of coastal wetland characteristics: the St. Clair Flats, Michigan. Proc. of the 13th Int. Symp. on Remote Sensing of the Environment. 1117-1129, Ann Arbor, Michigan.

Lyon, J. 1980. Data sources for Great Lakes wetlands. Proc. of the Annual Meeting Amer. Soc. Photogrammetry, St. Louis, MO., 516-528.

MDC 1982. Hazardous waste management in the Great Lakes Region: opportunities for economic development and resource recovery. Michigan Dept. of Commerce, Lansing, Report No. NBS-GCR-82-405. 412 p.

NRC and RSC 1985. The Great Lakes Water Quality Agreement. An evolving instrument for ecosystem management. National Research Council for the United States and The Royal Society of Canada. National Academy Press, Washington, D.C.

Phillips, L. 1984. Decision support for managers. In: H. Otway and M. Peltu (eds.), The Managerial Challenge of New Office Technology. Butterworths, London. 246p.

Reinert, R.E. 1970. Pesticide concentrations in Great Lakes fish. Pesticide Monitoring Journal, 3:233-240.

Schneider, J. 1984. JRC,ISPRA. personal communication.

Simon, H.A. 1960. The New Science of Management Decision. Harper and Row, New York.

Sonzogni, W.C. and Swain, W.R. 1982. Perspectives on U.S. Great Lakes chemical toxic substances research. Environmental Research Lab., MI.

Weigkricht, E. and Winkelbauer, L. 1986. Knowledge-based systems: overview and selected examples. WP-86-XX. International Institute for Applied Systems Analysis, A-2361 Laxenburg, Austria (in press).

Wierzbicki, A. 1983. A mathematical basis for satisfying decision making. Mathematical Modeling USA 3:391-405. (also appeared as ILASA RR-83-7)

Zhao, Ch., Winkelbauer, L. and Fedra, K. 1985. Advanced decision-oriented software for the management of hazardous substances. Part VI: The interactive decision-support module. CP-85-50, International Institute for Applied Systems Analysis, A-2361 Laxenburg, Austria. 39p.

CHAPTER 10

CLEANING ABANDONED CHEMICAL WASTE SITES AND THE LARGER
PROBLEM OF PREVENTING NEW DUMPS

Ross H. Hall

Pollution Probe Foundation, Toronto, Ontario, Canada
and McMaster University, Faculty of Health Sciences,
Hamilton, Ontario, Canada

What is the half-life of a human being - 35-40 years?
The chemical dumps that ring the Great Lakes were
created mostly within the span of half a human life.
The area studied by the author specifically, Niagara
Falls, NY. contains some 215 chemical waste dumps
within walking distance of the Falls.   That is, 215
dumps the size of Love Canal.   These dumps hold
collectively about eight million tons of toxic
chemicals whose environmental persistence is measured
in hundreds of years.   Waste dumping indeed has been a
by-product of the chemical industry since its
inception, but most of that eight million tons has been
dumped in the last 35-40 years: one reason, chemical
production has expanded about 50-fold in that period.

Project ahead one human half-life.   On-going chemical
production occurs at the same 50-fold rate.   Even if
there is no further increase in the rate of production,
the potential for creation of another 215 dumps the
size of Love Canal is palpable.   Niagara Falls is only
one of several industrial regions girdling the Great
Lakes.

The author makes this projection because today's
quality of Great Lakes water menaces the well-being of
the ecosystem, including its human inhabitants.
Evidence presented at this conference supports this
conclusion.   What outlook awaits us in the year 2020?

The half-life of a human embodies an ecological concept that frames chemical management. Many of the harmful effects of human exposure to toxic chemicals take 35 or more years to manifest themselves. The management decision taken today therefore, can have significant positive effects for the human population, and thus for the Great Lakes ecosystem. The author projects ahead to the year 2020 because the problem of chemical waste dumps has to be placed in the larger perspective of on-going chemical production and waste generation. We have to learn how to manage chemicals - not just chemical waste. That is, chemicals must be managed, starting from their production through to disposal in a way that reduces waste to zero. The need for comprehensive management is underscored by focussing on a legacy of the past 35 years, the chemical waste dumps of Niagara Falls. These dumps lie on top of a bedrock that is fractured horizontally and vertically. Their contents ooze downwards and outwards. The toxicants leach constantly into the ground water and thence Lake Ontario. This lake provides recreational and drinking water for six million people. The pressure for remediation is strong.

The United States Environmental Protection Agency (EPA) proposes to build containment walls around each dump. The walls stop outward oozing, directing the leaching downwards where the chemical-laden liquid will be collected and pumped out. The dumps will continue to ooze and leach for several human generations. The cost will be on-going both in money and vigilance to ensure that collection and pumping blocks movement of the toxic contents into the ground water.

Although the initial costs of building the containment walls and leachate collections systems are modest. An economist with Environment Canada has estimated that the long-term maintenance and environmental costs will raise the total cost well above that of the permanent solution of excavating and destroying the contents of the dumps.

Pollution Probe Foundation, together with Environmental Defense Fund (Washington, DC), organized in November, 1985 a conference on permanent solutions to the Niagara Falls dumps. Highlights are as follows: engineers from the U.S., Canada and Europe outlined existing techniques that work: excavation, purging and destruction by incineration or chemical treatment of the toxic wastes. The costs can be high, $500-$1000 a ton. But if those costs are amortized over the space of one half a human life, they need not seem exorbitant.

Inaction prevails however as leaders of the impacted
jurisdictions in the United States and Canada argue
over projected costs and effectiveness of containment
versus excavation/destruction.   And who should pay?
The industries which created the dumps or the
taxpayers? To reduce this time-wasting palaver a start
should be made immediately on the excavation and
destruction of at least one of the 215 dumps.   The
project would be a hands-on learning experience,
demonstrating actual costs and providing opportunity to
innovate cost-lowering techniques.

As stated at the outset of this paper, cleaning up of
chemical waste dumps eliminates an important source of
contamination created in the past.   But let us return
to the on-going management of chemical manufacture and
use so that within the space of the next human half-
life, environmental waste bottoms at zero.   Can we
achieve such perfection?

Other speakers at this conference urge a systematic
approach to dealing with environmental contamination.
Perhaps they too observe the fractured policy-making
jurisdictions, the disconnected research programs, the
uncoordinated management agencies.   All adding up to
ineffectual results and wasted talent - and continuing
pollution of the Great Lakes.

Two proposals for a systematic approach to the
management of chemicals are offered.   They are best
posed as two questions:

   1. What goals are we trying to achieve?

   2. What information do we need on which to
   formulate action policies to achieve these
   goals and what are the barriers to obtaining
   this information?

The Goal: Zero discharge of chemicals into the Great
Lakes, attained not by shutting down chemical industry,
but by managing aggressively chemical manufacture,
chemical use and chemical disposal within the
continuing economic and social evolution of our
society. This goal cancels the view of the Great Lakes
as a sink for chemical waste.   It cancels the
unanswerable question of how much chemical loading the
lakes can withstand.   It cancels argument over the
definition of toxic: what is and what is not a toxic
chemical. Zero discharge applies to all chemicals.

To achieve this goal chemicals need to be managed at
three stages: manufacture, use and disposal.   At no
time is a chemical invisible to management.   If a

chemical cannot be used or disposed of without contaminating the environment, the manufacture of that chemical would cease. Elements of this management plan already exist. The managers of many chemical companies voluntarily abort the manufacture of any new chemical if they foresee a waste problem. The Toxic Substances Control Act (TSCA) in the United States and the Environmental Contaminants Act in Canada, embody the idea of control of chemicals at the point of manufacture or importation.

These two acts suffer, however, from exaggerated attention devoted to the definition of toxicity for each chemical manufactured and used in commerce. Attempted classification of each chemical according to its degree of toxicity attracts endless dispute because the limited scientific ability to determine chemical toxicity leaves too much uncertainty.

A goal of zero discharge eliminates dispute and focuses on an easily measured question for all chemicals: Is there leakage of the chemical into the environment?

The Information: Systematic management of chemicals as proposed, requires new types of information. Much technical research data now generated and much current policy-making support only micro-management. That is, management within one geographical region or management of a shard of our industrial society.

The concept of zero discharge has to be built into chemical industry, requiring new engineering and design. The added cost can be zero. The 3M Company for example, with its in-house policy of zero waste has demonstrated that pollution prevention pays.

An information base for systematic management needs also to integrate technical issues with social and economic planning. For example, if obstinate chemicals have to be phased out of production, any economic and social dislocation has to be researched and appropriately met in advance.

The body of environment laws in both Canada and the United States is fragmented into single media or single jurisdictions. Separate laws exist for air, water, land, agriculture, forestry, industrial plants, etc.. New laws are required that acknowledge the multi-media movement of chemicals, including the medium of human bodies, and that promote systematically the manufacture. use and disposal of chemicals with zero discharge.

Zero discharge requires on-going regulation. Existing
regulatory bodies suffer weak mandates or lack of will
to enforce zero discharge. Having set a goal of zero
discharge into the environment, we need now to research
the most appropriate ways to regulate or encourage
achievement of that goal.

In brief, it is easy to speak of a systematic approach
to enhancing the quality of the Great Lakes
environment. The author does not believe that a
systematic approach will gel. However, without
fundamentally reorganizing our scientific and social
research, our environmental laws and our decision-
making regulatory agencies. Let's do it before another
human half-life passes by.

CHAPTER 11

THE 3P PROGRAM: AN EFFECTIVE APPROACH TO INDUSTRIAL POLLUTION

Joseph T. Ling

Executive Consultant, 3M-3M Center, St. Paul Minnesota 55144, USA

ABSTRACT

The best way to control toxic substances from industry is at the source. Since industrial pollution is a visible sign of inefficient use of resources, 3M developed a program to fight pollution by not creating it. Industry's traditional approach has been the use of add-on control equipment that changes the form of pollution but does not eliminate it. The 3M Pollution Prevention Pays (3P) program eliminates or reduces pollutants, conserves resources, and encourages innovative technology through product reformulation, process changes, equipment redesign, and recycling or reuse of process waste. Started in 1975, the 3P program, involving 3M operations in the United States and 22 other countries and annually prevents more than 40,000 tons of pollutants and 1.6 billion gallons of wastewater. 3P savings to date total $300 million. 3M's prevention approach has had national and international impact. The program has received awards from the U.S. Environmental Protection Agency and other organizations. Some states have adopted pollution prevention as environmental policy. Several world organizations, including the United Nations Environmental Programme and the Organization for Economic Cooperation and Development, have endorsed and promoted the concept. Pollution prevention has become government policy in several countries, including

France and Britain. A number of major industrial companies also started prevention programs. Industry, however, is only one source of pollution. Other sources also must be effectively addressed. Since many environmental concerns are international in scope, meaningful government incentives and expanded international cooperation are vital to the development and implementation of innovative solutions to environmental problems, including those of the large lakes.

INTRODUCTION

It is a personal and professional pleasure to participate in this conference because protection of the environment has been my occupation and preoccupation for more than 40 years. I also welcome every opportunity to meet with those who share these concerns and seek responsible solutions to these problems.

The concern that brings this distinguished group together is preservation of the world's large lakes. The problem is to save them from pollution. The solution requires the thoughtful consideration, planning and cooperation of those in science, technology, industry, and government, working together for the public interest - which involves all of us.

Today, I submit for your consideration and commend for your support a program that can be part of the solution. It protects the environment from pollution, especially industriall sources, and at the same time saves resources and money, and also encourages the development and implementation of new and innovative technology.

The program has done these things and more for the environment and 3M, the company I had the privilege to serve during the past 25 years. It fits. or has the potential to fit, almost anywhere where society's standard of living is being upgraded through industrial output and its environment downgraded by the pollution created.

This program is called **Pollution Prevention Pays - 3 P** for short.

WHY 3P STARTED

The basic concept of 3P is that most pollutants are unused resources due to the inefficiency of manufacturing processes and that pollutants can be converted to valuable resources with innovative techniques.

3M started the 3P program in 1975 at a time when industry faced a torrent of new rules and regulations to improve and protect the environment. Compliance was a costly mandate. The concurrent energy crisis and a downturn in the economy also contributed to the need for a careful evaluation of ways to meet the compliance mandate.

The first to be examined was pollution control, the conventional wisdom then, and still the dominant approach to abatement of industriall pollution. This is the so-called "black box" method - attaching an add-on device at the end of the manufacturing process to capture pollution <u>after</u> it is created.

This approach has many limitations. Natural resources, energy and manpower, and energy are consumed to build the black box and more resources continually consumed every year to operate it. Furthermore, this type of control is, in truth, merely a repackaging of pollution. It only temporarily contains the pollutant but does not eliminate it, for residue from the black box represents another form of <u>on-site</u> pollution and poses a complicated and expensive disposal problem of its own. It should be recognized that it takes more resources to dispose of this residue, and the disposal produces more pollution.

In addition, using the black box approach contributes to another type of pollution, which is generated by those who supply the material and energy to build and operate the black box. This pollution could be generated at facilities many kilometers from the black box, and we call it <u>off-site</u> pollution.

Both <u>on-site</u> and <u>off-site</u> pollution are generated from the factories which represent the **first** generation of industrial pollution. However, industry must also address the pollution from the use of its products. This is **second** generation pollution for industry and is global in nature. A product made by a company in Michigan may be used as a raw material in France, Japan, or California. In short order, the second generation problem of the Michigan company becomes a first generation problem for the company in France.

Further, product disposal becomes a major concern in
many countries. Examples are lead and mercury from
batteries, PCB in transformers, phosphates in
detergents, and chloro chemicals in insecticides. This
represents a **third** generation pollution problem for
industry.

Obviously, the black box approach at best will only
take care of the first generation problem and never
have any effect on the second and third generation
problems.

However, industry must have a solution for all three
generations of pollution if it is to fulfill its
environmental responsibility.

In the final analysis, 3M concluded that pollution
should be controlled, eliminated, or minimized at the
source.

IMPLEMENTING 3P

Preventing pollution at the source is not a new idea,
but 3M was the first major multinational company to
make it standard world-wide operating procedure.

Implementing the 3P program is a formidable task in a
company with more than 45 major product lines and more
than 80,000 employees on six continents. The following
basic approaches have been used:

1. Product reformulation

2. Manufacturing process modification

3. Equipment redesign

4. Recycling and reusing process wastes

All employees are encouraged to seek and submit
preventive solutions. The criteria of acceptance
include environmental enhancement, economic benefits
and technical innovation. All submittals are evaluated
by a committee composed of representatives of research,
manufacturing, engineering, and other concerned staff
organizations. Employee participants in an approved
program receive plaques and certificates and are
recognized by top management at appropriate ceremonial
occasions. 3P achievement is also noted in an
employee's file and can be a factor in career
enhancement.

Here are two brief examples of 3P success stories. One
involved replacing two hazardous stabilizers in Static
Control Mats used under personal computers. The change
allowed for more than $100,000 of continued sales and
eliminated 700 tons per year of hazardous wastes. The
second case was use of a sonic spray cleaning system to
replace solvents in a chemical manufacturing process.
This resulted in a first-year saving of approximately
$600,000 at a modification cost of only $36,000 and the
elimination of 1,000 tons per year of water pollutants.

3P ACCOMPLISHMENTS

3P has produced excellent results during the past 11
years. The program now annually eliminates 110,000
tons of air pollutants, 13,200 tons of water
pollutants, 275,000 tons of solid waste and sludge, and
more than 1.5 billion gallons of wastewater. These are
cumulative first-year results only. They do not
reflect the benefit of second and later-year results.

Economic returns to date total more than $300 million,
and this represents only the results of the first-year
savings. Components of this saving include pollution
control equipment that did not have to be constructed,
savings in raw materials, energy and other operating
costs, and retained sales from products reformulated to
remove a pollutant. Eighty percent of the savings are
from U.S. operations, the balance from 3M international
operations involving 22 countries on six continents.

To date 1,900 projects have won approval in the 3P
program, involving contributions from more than 1,500
employees.

The pollution prevention concept is a plank in 3M's
corporate environmental policy.

IMPACT OF 3P

3P has had a significant favorable impact on
environmental practice and policy world-wide.

In the United States

In the United States, the Environmental Protection
Agency (EPA) endorsed the 3P concept in the early days
of the program. The agency and the U.S. Department of
Commerce sponsored a series of four national
conferences in 1977 around the 3P approach. They were
held in Chicago, Boston, Dallas, and San Francisco.
The 3M chairman and several members of the 3M
management team and the author spoke at these
conferences, underscoring management's support of the
3P program, a commitment that is as strong today as it
was then.

3M has shared its pollution prevention experience
widely with American industry and with governmental
agencies. As a result, many companies developed and
implemented their own pollution prevention efforts
along the lines of 3P. Since 3P was started 11 years
ago, there has been a steady stream of inquiries from
companies, government agencies, and conservation
organizations and individuals.

Several states, including North Carolina, Alabama,
Tennessee, and Minnesota, have adopted the pollution
prevention pays concept in their environmental
programs. The State of Alabama recently sponsored a
conference around that theme, and Kentucky plans a
similar meeting this year.

The EPA and several state governments have honored 3M
with awards for 3P program efforts.

International

3P has also been well accepted outside the United
States. 3M first reported its new pollution abatement
philosophy at a meeting of the Economic Commission for
Europe in Paris in 1976. Since then, 3P has been
introduced to and endorsed by a number of other
international organizations.

At the request of the United Nations Environmental
Program (UNEP), 3M in 1982 prepared a monograph
entitled, **"Low or Non-Pollution Technology Through
Pollution Prevention."** This was in recognition of the
tenth aniversary of the first World Conference on Human
Environment in Stockholm. More than 15,000 copies of
the monograph have been distributed worldwide.

With a grant from the U.S State Department, the World Environment Center, New York City, published an updated edition of this publication in 1985 and translated it into French for the benefit of the French-speaking countries in Africa.

The 69-nation World Industry Conference on Environmental Management, held in France in 1984, strongly endorsed the 3P concept. The conference, co-sponsored by UNEP and the International Chamber of Commerce, declared that,

> ..."An anticipatory and preventive approach to the threat of environmental degradation is preferable to correcting environmental problems after they have occurred."

A similar endorsement was declared by the Organization for Economic Cooperation and Development (OECD), comprising 24 developed nations, and the European Economic Community (Common Market). The preventive approach was also adopted by the Governing Council, UNEP's top policy body, in Nairobi, Kenya, last year.

Great Britain, France, Australia and Mexico, among other countries, have adopted pollution prevention as part of their official environmental policy.

The World Environment Center in 1985 honored 3P with the first International Corporate Environmental Achievement Award in Washington, DC.

CONCLUSIONS

The 3P approach to industrial pollution works, as the 3M experience has shown. It can be applied successfully by many others in industry. In the long run, prevention at the source is the most effective way to spare our water, air, and land from toxics and other pollutants. It contributes to a cleaner environment, better use of resources, advances in technology, and saves money.

However, industry is only one source of pollution. To enhance and maintain desirable environmental quality, other pollution sources must also be abated. For example, non-point sources, such as surface runoff from cultivated crop land, have not yet been properly addressed.

Obviously, industry must do its share, but government must also help if pollution prevention is to play an expanding role in world environmental management. Environmental policies and regulations must encourage technical innovation and set pollution abatement goals that are in accord with technological and financial capabilities and realities. Government can incorporate appropriate economic instruments into environmental policy to help induce innovative, preventive technology.

Since many large lakes touch the borders of several countries, international cooperation is an indispensable component of effective environmental policy.

The United States and Canada are an example of international cooperation. Through the International Joint Commission, they work to preserve and enhance the quality of this Great Lakes region. It is to be hoped that this conference will generate a greater spirit of international cooperation in dealing with the environmental issues related to large lakes around the world.

Our greatest natural treasure is the environment: water, air, land and resources. It belongs to all of us and will require the efforts and cooperation of all to protect it from the degradation of pollution.

The best protection against a problem is to remove its causes. That is why the pollution prevention approach is the most effective long-term environmental strategy. It really pays.

A wise man said:..."**An ounce of prevention is worth a pound of cure....** That is the challenge and opportunity for us in the environmental field today.

CHAPTER 12

INDUSTRIAL SITING IN A DEVELOPING COUNTRY THE CASE OF
ARGENTINA - Its Context - The Environmental Dimension,
Hazardous Waste and Large Lakes

Pedro Tarak

Indiana University, Fondacion Ambiente y Recursos
Naturales, Buenos Aires, Argentina

ABSTRACT

The context within which an industrial siting process
takes place in most developing countries is
characterized by the lack of environmental legal
constraints. Moreover, most of the developing
countries do not have a genuine (self-made)
environmental policy whereby people can determine the
kind of environment they wish to have. Neither, do
they practice standardized environmental impact
assessment mechanisms nor land use planning as possible
tools for effective environmental management. This
general picture also includes the fact that developing
countries do not have a specific-sectorial management
of hazardous waste, thus dealing with them together
with other kinds of pollutants. Also, the ecosystem
management approach for large lakes is absolutely
ignored in the decision-making processes connected with
them.

In this context, industrial siting becomes exclusively
an administrative process where environmental aspects
stay mostly under the discretional spheres of the
officers in charge of the permit granting process.
These officers are governed by the internal regulations
of the Administration. Moreover, and very often,
environmental concerns are neutralized by not only

economic and social benefits of the projects, but also by behavioral patterns of societies and its bureaucracies that allow corruptive practices such as briberies, personal and political favouritisms and authoritarian threats to succeed.

This paper describes the different systems where industrial siting can take place, the kind of policies that should be promoted to avoid imitating the developing world's political-historical evolution and the various mechanisms available which can be used to introduce direct public involvement during industrial siting processes.

Finally, a case study is analyzed regarding the design of a new "Technological Park" in the area of Lake Nahuel Huapi, Argentina.

INTRODUCTION

The purpose of this paper is to outline the kind of limitation and lack of policy and legal constraints that characterize an industrial siting process in a developing country like Argentina, when attempting to consider environmental factors or adequate hazardous waste management. However, it is important to analyze a few concepts widely used in developed societies, but which do not very adequately apply to developing countries:

  1. What determines the degree of development of a society is not the level of technology, Gross National Product or per capita income (or any other abstract standard that can be specifically defined), but its capacity to maximize the energetic base of the society, minimize entropy and achieve social productivity for the sake of the satisfaction of material and spiritual needs determined by each particular culture.

  Cano (1985), citing Rifkin (1981), explains that the energetic base is conformed by energy elements on which a civilization, in a given time and space, will determine and design its public and private institutions. Cano adds that "...if mankind wishes to protect itself (within an environmental crisis situation) it should redesign its political and administrative institutions with simple patterns, generating less

entropy".   In this context, one of the
critical ingredients lacking in most of the
developing   countries   is   the   kind   of
organization, both public and private, that
may enable societies to minimize entropy and
achieve social productivity available for a
development   process.   However,   such
organization   (and   also   the   technology
necessary for its operation) must necessarily
be responsive to the cultural evolution of
each society.   Thus, it is important to avoid
thinking that "exports" of organizational
models, accompanied by technologies adequate
for these models, will guarantee development
in other latitudes.

2. Development is a continuing process and
not a permanent stage.   For Sunkel (1981) it
is the "...transformation process of the
natural environment into an artificially and
specialized   built   milieu."   This
transformation   ..."is   achieved   by   the
interaction of four fundamental elements:
energy, technology, social organization and
culture"...(Sunkel and Leal 1984).

As a result, each society will define its own
"development   style".   In   addition,   each
society will determine its own environmental
policy, which in turn is the end-product of
the   interaction   of   four   other   elements:
science,   technology,   economics   and
culture (Tarak   1986),   applied   to   a
..."permanent definition of a desirable and
possible environment wanted by a given
society in a given time and space"...(Tarak
1986).   The environmental policy process is
therefore unique to every society mostly
because the cultural element is always
exclusive of each society (Caldwell 1984).
For this reason it is important to avoid
"exports" of environmental policies or their
legislative   expression   and   their
institutional responses, because such an
approach means mostly the application of the
political will (regarding the kind of
environment that a society wants to have) of
one country to another.   In the context of
South - North relationships, this process
aggravates domination linkages and inhibits
autonomous   decision   making   processes,
responsive   to   genuine   social   and
environmental needs in the South.   Economic
domination relationships also arise regarding

necessary scientific research and technology for the application of imported environmental policies.

For an analysis of the Industrial Siting Processes in developing nations, from a methodology standpoint, it is conceptually wrong to elaborate on generalized situational descriptions applicable to any country of the Third World. Probably, such methodology has been used across the board in the educational programs of both North and South. Yet, the shortcomings of such training, simplifying reality to our own mental capacities, has produced a number of mistakes in international political and economic relations. Therefore, this paper will attempt to analyze only one country-case, Argentina, and it will depend on the reader if he wishes to extrapolate Argentine information and experience to other societies.

Although they are not going to be analyzed here, several characteristics, regarding the context within which an industrial siting process takes place in developing nations, might operate currently in all countries. Some of these are:

1. limited data on social, economic and environmental factors. Varying its degree, countries might have a better profile on their social and economic needs than on environmental constraints and capacities;

2. limited availability of financial resources for adequate government assessment on the benefits of the industrial project;

3. limited number of trained officers, with adequate background in charge of industrial siting processes, capable of balancing short-term economic and social benefits with long-term environmental productivity;

4. the need to satisfy immediate basic demands such as food, health, clothing and shelter allotting all available financial resources for such purposes, ignoring most of the time the environmental dimension;

5. political and economic pressures from dominating interest groups, many times based in dominating countries, that neutralize all attempts for environmental or even social considerations.

Another aspect which will not be dealt with in this paper is referred to administrative procedures required by the different existing legal systems which are in the developing world or more specifically - in Argentina. This is purposely omitted because most of the procedural systems applied in the South for industrial siting processes have had their origin in systems from the developed world. This is an obvious consequence of colonial times.

CONTEXT IN WHICH AN INDUSTRIAL SITING PROCESS TAKES PLACE

Lack of Environmental Policies and Law

If environmental policy is the political process whereby a society permanently defines the kind of environment that it wishes to have, in a given time and space (Tarak 1986), it can be affirmed that Argentina has not yet experienced such policy making process as a political practice (Tarak 1986). Instead, sectorial policies on natural resources (energy, oil and gas, nuclear, agriculture, etc) have been strongly developed trying to maximize utilities. In the field of water quality protection, a few states and municipalities have applied effluent standards to control water pollution (Moyano 1986). However, these standards have had no connection with water quality goals, set by the communities of those municipalities or states, nor do they respond to social determinations on possible uses of watercourses - human consumption, recreation, navigation, urban sewage degradation, industrial use, etc. (Castignino 1986). In practice, effluent standards have been "imported" from developed countries (generally, from the U.S.A.) and then applied locally (at least theoretically speaking). Conceptually, what this means is that political environmental goals of other societies, goals which have been broadly discussed and measured by others, who live under different social, economic and environmental circumstances, are adopted for some Argentine communities by its technical bureaucracies. This approach establishes the same kind of treatment requirements as in those countries where goals were previously set. But in reality, polluters keep on polluting and water bodies become more and more degraded. Other factors neutralize the few existing legal requirements. In the area of Buenos Aires, the polluter-pays principle was applied to the industry that discharged its effluents to watercourses used by the National Sanitary Works for human consumption (rivers Plate, Reconquista, Riachuelo, etc.) (National Decree 1978). However, the effluent-charge system, institutionalized by Decree 2125/78, only designed a matrix for the quantification of the due amounts by each polluter. Funds so obtained were not utilized for wastewater treatment purposes by the government so the system only operated as a taxing mechanism. The Decree was revoked by a Federal Court on August 16, 1984, declaring the polluter-charge system unconstitutional (Court Decision 1985).

The lack of environmental policy and law making processes has caused National Senator Miguel Mathus Escorihuela, to submit a bill to Congress creating the State Secretary for Environmental Policy (Mathus Escorihuela 1984). Its functions are:

    1.the permanent assessment of the environment,

    2.the permanent proposal of environmental quality goals, and

    3.to guarantee that environmental factors will be taken into account by the Federal agencies during their administrative decision making processes.

The bill was submitted October 1984. However, it has not yet been discussed either by the Senate or even by its Committee on Natural Resources and Human Environment. Such political behaviour explains the lack of genuine interest by the average legislator, by the partisans of the different political parties represented in the chambers and by the general public that does not feel a critical need to activate its government bodies in an environmental policy formulation process (Caldwell 1984).

Lack of Toxic Waste Policies and Laws

In a context of lacking environmental policies and legislation, hazardous waste policy and law are nonexistent. Argentine law does not distinguish toxic waste from other kinds of pollutants. Its management has had no differentiation from other industrial waste. Thus, regarding toxic waste (persistent, corrosive, bioaccumulate, etc.), these are discharged permanently into watercourses or disposed of in "unclassified" open dumps either next to the generator's premises or in unknown areas selected by truck-drivers hired to collect the waste. The only legal limitation that a generator has is the prohibition to discharge "poisons" that directly affect public health, as defined by the Criminal Code (usually these poisons are those prepared purposely as such). The dangers of toxic waste in the environment have been the subject of conferences, articles in newspapers, radio and TV interviews during the last two years. Slowly, Argentine society is beginning to express its concern and, also political will to define a policy towards the problem (Jesses 1985). Simultaneously, a government company, in charge of the area of Greater Buenos Aires, has called a bid for the operation of the first hazardous waste treatment and disposal plant in Latin America (Tarak 1985). Hopefully, the moment the plant is built, the legislative bodies will enact the appropriate legislation, obliging generators of hazardous waste to avoid every kind of discharge into the environment, thereby preventing the irreversible effects to public health and environmental systems.

Lack of Regional Land Use Plans and Zoning as Tools for Adequate Environmental Management

Most of the municipalities in Argentina have municipal planning codes which regulate the use of space, determine the kind of activities authorized to take place in specific areas (residential, industrial, recreational, etc.) or define specifics concerning building of structures, public services, etc. However, very few have planning laws that consider environmental restrictions for the area (Municipality of Buenos Aires 1985). The Municipality of El Bolson (Province of Rio Negro) has declared its entire jurisdiction a "nuclear free zone", prohibiting any nuclear activity in the area. Yet, such prohibition responds more to the need to regulate an activity than to use land in accordance with human needs and environmental restrictions.

Regarding regional land use plans, the Province of Buenos Aires is the only state with a specific law (Caldwell 1984). Municipalities are required to follow land use criteria as established in the Act. Such regional planning enables coordination among municipalities. Other provincial or national laws establish sectorial resource management legislation with conservation units (reservoirs, protected areas, etc.) separating some areas for those purposes. However, these do not mesh into broader regional land use planning laws.

Few Standardized Environmental Impact Assessment Processes Country-Wide

Environmental impact assessment (EIA) as a social practice to avoid the ..."consequences of the failure of human insight and ingenuity to predict and prevent the ill effects of human imagination and purpose"... (Caldwell 1984), has only recently been introduced to Argentina in the Province of Cordoba (Province of Cordoba 1985). Instead, until now, environmental impact assessment has only taken place occasionally during the planning stages of large public projects, such as hydroelectric dams (i.e. Salto Grande) (Cravitto 1984) where also land use planning has been used for territorial management of affected areas. At the moment, several state parliaments and a few municipalities are analyzing the feasibility to incorporate EIA into administrative permit granting procedures and certain legislative proposals. This then will incorporate the environmental dimension into the mental structure of political and administrative decision makers.

Major Discretionality of the Officers in Charge of the Industrial Siting Process

The combination of the four factors cited and analyzed above, allow a great manouvering possibility to any officer in charge of the industrial siting process. The lack of environmental laws establishing quality parameters that can determine the characteristics of industrial settlements, the lack of hazardous waste laws different from any other pollution-control mechanism, the lack of land use mandates and environmental assessment as a current procedural practice, and above all, the lack of a social-political environmental policy formulation leave the authorities with major discretional powers in considering environmental elements when granting permits, both for siting or for operation of industrial facilities.

SYNOPSIS OF THE INDUSTRIAL SITING PROCESS IN ARGENTINA

Some Factual Considerations

Based on the constitutional right to develop any legitimate industrial activity (Art. 14) any inhabitant of Argentina may settle his industry in the territory of the nation.   In principle, municipalities are in charge of the permit granting process (Vanosi 1984, Dromi 1983).   Usually, under the present economic situation, any municipality is eager to attract industries to its jurisdiction (more jobs, more income from taxes, etc.). Few environmental restrictions are bound to be imposed by the municipalities.   However, municipalities do not have major fiscal tools to attract industry.   For this reason provinces and the federal government have designed industrial promotion systems, implemented by either state or national laws (Federal Law 1983).   These usually allow tax exemptions, long-term low interest rate, subsidies, trade barrier exemptions, etc.. These in turn profit many industrial ventures.   Most of the "industrial promotions laws" require that industries favoured by these laws meet environmental parameters, that are later defined (on a case by case analysis) by the promotion granting officer, prior to an opinion by the technical environmental agency.

In a context of great discretionality of the authorities, negotiations take place with each applicant regarding the construction of pollution abatement facilities or devices. An agreement concerning environmental matters is always obtained. In fact, the National Secretary of Industry has never rejected an industrial project because of environmental factors.

For the new industry, there are three siting alternatives:

1. Siting in an "industrial zone"

2. Siting in an "industrial park"

3. Siting in a "technological park"

The first alternative is most common.   In an **"industrial zone "** small and medium sized industry establish in an area defined as such by a zoning ordinance.

These areas are usually considering proximity to urban settlements (convenient for transportation of personnel), proximity to sewerage systems, proximity with watercourses to be utilized as industrial waste recipients, and several other variables that take into account economic and public health considerations.

The **"industrial park"** areas are a novel system. Here a group of industries are sited within a specific area sharing common services, among which water pollution control facilities are found. Industrial parks can either be owned by governments (state or federal), by private companies or by users cooperatives. In all cases, the siting of industrial parks is also the final result of an administrative process.

Finally, the **"technological park"** is also an industrial area where industries that belong to them need intensive technological services to operate, such as computer services, highly sophisticated laboratories, etc.. Technological parks (only two are in the process of being designed) are planned with industries such as biotechnology, software production, robotics, etc.. These industrial models are only planned where highly qualified human resources are concentrated in the area but do not have technological tools to maximize their productivity. The City of Bariloche, next to the Nahuel Huapi Lake, is now trying to design its own technological park.

Sociological Context

Regarding advertisement of an industrial siting application, and from the ordinary citizen's perspective, information is not systematically delivered to neighbours or to the general public to solicit opinions (positive or negative) concerning the project. Usually, such information is obtained through political and friendship connections. It should be added that Argentine law does not grant citizen rights to access to information (Mitchell 1980). Coincidentally, it also does not grant the right to participate in the industrial siting process, through public hearings or public consultations (enquete publique) (Tarak 1984). Nor does it grant the right to sue to individual citizens or citizen groups who wish to question an administrative decision (even if they are not affected in their property or person) (Tarak 1983). For example, an industrial siting permit that does not consider environmental factors. Within a political system that does not enable citizens to become potential controllers of decision-making processes, the final decision is probably bound to be influenced by social corruptive mechanisms:

1. Bribes. The payment of cash money or other benefits to influence decision-makers in municipalities, state executives or officers in the federal government is standard practice. Both, for the siting authorization and the operation of an industrial plant, permit granting officers and inspectors are bound to be "influenced" with some kind of bribery. This practice is so current that many companies include this factor in their budgetary calculations.

2. Payment of Political Favours or Friendship Deals. Many times political parties or government authorities need to pay back favours to industries that need to operate at a specific site (i.e. for economic reasons). The present structure of the political parties allow, very much so, to exert influence on the administrators who have hierarchical relationships with members of the party. The same applies to friendship linkages.

3. Authoritarian Ingredients in the Hierarchical Structure of the Public Administration. In many occasions a lower officer (most of the times this is the case of environmental officers) is threatened by authorities in higher position so as to silence an adverse opinion. This can happen when balancing environmental considerations with economic or other factors. This factor is perhaps very common in a society like Argentina's that has institutionalized authoritarianism with both democratic and military political systems.

CONCLUSIONS

The combination of factors such as:

1. great discretionality of the authorities to assess environmental values or consider water quality,

2.the   existence   of   social   corruptive
practices, and

3.lack of citizen controls to administrative
decisions

leaves most of the fate of water bodies (aquifers,
rivers and lakes) to the good will of officers in
charge of those decisions. However, with the return to
democracy in Argentina, public participation has
rapidly increased in pace and intensity. Environmental
non-governmental organizations have massively appeared
and constantly influence decision-makers through the
media, lobbying politicians, organizing public meetings
and building up general awareness for environmental
protection. A few parliamentarians have started to
call for public hearings to discuss different bills and
ideas with private citizens and non-governmental
organizations. Also several judges have granted rights
to sue to ordinary citizens concerned about
environmental problems. State legislatures and both
chambers of the National Congress are permanently
dealing with sectorial environmental bills in their
specialized committees. Obviously, these trends show
that a critical moment has arrived for a structural
change of the society. Most probably the picture will
be different in the near future. The present
administration has already launched a government
program for the Second Republic, completing the
traditional democracy with a participative system.
Evidently, the context of industrial siting changes
radically with a participative democracy where people
can actively control public administration.

RECOMMENDATIONS

There are two major recommendations which can be made
at this time:

    1.   Water quality and toxic waste policies
    2.   More public involvement

Water Quality and Toxic Waste Policies

Considering economic limitations through which
Argentina lives constantly (foreign debt, trade
barriers in traditional markets, industrial recession,
unemployment, etc.), the country should adopt in the
short term a hazardous waste policy rather than a water

quality policy.   This strategy is totally different than that adopted historically by the developed countries, which has already produced massive technology for water pollution control plants and is now exporting it to the developing world, with the support of the international banking systems.   Instead, a hazardous waste policy allows the elimination, from the very beginning, pollutants which have irreversible effects on the environment and public health.   Its implementation is also significantly cheaper than a water pollution control program.   Whereas the latter requires the construction of numerous industrial and municipal treatment plants, with huge administrative control systems, the former needs a few hazardous waste treatment and final disposal plants, with a smaller administrative cost for government control, and an easier way to determine generators and follow the steps taken to isolate the hazardous waste (cradle to grave criteria).   This strategy does not mean that industrial or municipal effluent treatment plants should not be built.   It only signifies that in Argentina, where new technological processes are introduced constantly, a hazardous waste policy should precede other pollution control policies, especially in light of the scarcity of financial resources in the country.   In this respect it would be very beneficial if such a strategy is discussed by both the World Bank and the Interamerican Development Bank when defining financial resource allocation in the field of environmental protection.

More Public Involvement

Taking into account the social and political context in which an industrial siting process takes place, it is important to develop new mechanisms that enable more active public involvement in such a process. Advertisement of applications to municipalities, state and federal government, details of the project, impacts to the neighbourhood and the general environment, should be obligatory.   This will allow interested neighbours, ordinary citizens and environmental associations to be well informed in case they want to add information, make suggestions or object to the project.   Public hearings during the siting process of major projects should be common practice.   Finally, the right to sue should be granted by law to any citizen or environmental non-governmental organization as a last resort to participate as controlling elements of the existing bureaucracies.

A CASE HISTORY - LAKE NAHUEL HUAPI

Lake Nahuel Huapi is a large lake located in the north-western edge of the Andean Patagonic region, of approximately 570 square kilometers. It belongs to a larger hydrographic basin (Limay-rio Negro) where waters flow to the Atlantic Ocean. The lake is characterized by its very low biological productivity (no eutrophication), slow water renewal time, few associated coastal wetlands and limited diversity of animal, bird and fish species.

The lake is geographically located on the border of two Argentine provinces: Rio Negro and Neuquen. However, the only significant human settlement is that of the City of San Carlos de Bariloche, in the Province of Rio Negro. With a population of approximately 80,000 inhabitants plus an average of 20,000 tourists, Bariloche is a growing city that extends 40 kms along the lake. However, its major industry is tourism, with significant timber extraction activities.

The jurisdictional borders of Bariloche are surrounded by federal lands belonging to the National Parks system, that legally speaking, manages the entire coastline of the lake and its waters. Yet, the city discharges its municipal sewage into the lake without treatment and without agreement from the federal government. Samples taken in 1980, showed that coastal waters near the effluent discharge areas are not suitable for human consumption. This is mostly due to the high concentration of bacteria.

Apart from concentrated biological pollution in the coastal area near the city, in principle the lake is not affected by other kinds of pollutants (toxic waste is not present in the area). This is mostly due to the fact that the region has not developed activities that produce large scale pollution. In any case, very little data are available on water quality parameters. However, due to the population growth of Bariloche and the future federal district in the Patagonic city of Viedma, industrialization of the city is bound to occur. With this perspective, the city of Bariloche is planning a "technological park" only with "clean industry". The planners of the park base their decision on two factors:

> 1.Lake Nahuel Huapi belongs to the Park system and the water quality goal should always be the highest possible ("zero pollution" should be approached) considering that the main reason for the growth of the

city is the alternative "natural" life style
that if offers to Argentines of other
regions. This is in addition to tourism from
the country and abroad, and

2.the population of the city has a highly
qualified professional group that would
augment its working capacity if technological
services are available in the area. Such
concentration of "intelligence" is unique in
Latin America.

Regarding other industries, the City of Bariloche has a
Planning Code that regulates zoning areas.
Accordingly, existing industry will slowly be relocated
in the "industrial zone", for which effluent treatment
plants are to be designed. Outside this zone, a
general prohibition will not permit any other
industries to operate. The Province of Rio Negro does
not have a regional land use plan that can be applied
to Bariloche.

At this time the City council is analyzing a draft
ordinance adopting environmental impact assessment
procedures for municipal decisions. If enacted, it
would be the first municipality in the country with
such a practice. The local university, University of
Comahue, is now in the process of producing the
"environmental map" of the municipality. The purpose
of the "map" is to produce available data on
environmental factors, both of the entire municipal
jurisdictional area and of Lake Nahuel Huapi.

Finally, it should be noted that the concept of system-
management of large lakes has not yet been developed in
Argentina for application in policy and law making
processes. However, river basin units are being
focused for different purposes by several provinces
which, in turn, are bound to the mandates of
interprovincial agreements. Many times these
agreements create interjurisdictional basin agencies in
charge of the specific activity and purpose of the
treaty.

In the case of Lake Nahuel Huapi, the present and
future activities that will affect the system will
operate under several jurisdictions. Most assuredly,
an interjurisdictional agreement can serve the purpose
for an ecosystem approach to the lake (Meier 1982).

134   Pedro Tarak

REFERENCES

Caldwell, K.L. 1984. International Environmental Policy, Emergencies and Dimensions. Duke Press Policy Studies, Durham, North Carolina.

Cano, G. J. 1985. Dissertation presented at the occasion of his inaugoration into the National Academy of Moral and Political Sciences, Argentina, May 22, 1985.

Cano, G. J. 1984. La participacion popular en la gestion ambiental. Ambiente y Recursos Naturales, Vol.I, No.2, 1984.

Castagnino, W. 1986. Criterios para el control de contaminacion del agua a nivel regional. In: El Principio Contaminador. (ed.) International Commission for Environmental Law and Administration. Fraterna, Buenos Aires.

Craviotto, M.A. 1984. Metodologia para el ordenamiento del espacio y desarrollo ambiental en el perilago argentino de Salto Grande. Ambiente y Recursos Naturales, Vol.I, No.1, 1984, p.34.

Court Decison 1984. Published in Ambiente y Recursos Naturales, Vol.II, No.1, 1985, p.75

Dromi, J.R. 1983. Federalismo y Municipio. Ciudad Argentina, 1983, Mendoza, Argentina.

Federal Law 1983. Law No.21608/77 amended by Act 22876/83; Laws 22021/79, 22702/82, 22973/83.

Interview J.A. Jesses 1985. Cuadernos de Ambientalismo. Ambiente y Recursos Naturales, Vol.II, No.7, 1985, p.30

Mathus Escorihuela, M. 1984. Proyecto de ley sobre 'Creacion de la Secretaria de Estado de Politica Ambiental'. Ambiente y Recursos Naturales, Vol.I, No.4, 1984, p.95

Meier, E.H. 1982. Estudios de derecho y administracion del ambiente y de los recursos naturales renovables. Part VII. Un caso especial: la conservacion, defensa y mejoramiento de la cuenca hidrografica del Lago de Valencia. Artee, Caracas, 1982.

Moyano, A. 1986.   Cuestion juridica interjurisdiccional del suministro de agua y del saneamiento ambiental. Agua - Journal for Technology, Treatment and Water Sanitation. Vol.X, No.44, 1986.

Mitchell, H. 1980.   Access to information and policy making: acomparative study.   Ontario Commission on Freedom of Information and Individual Privacy. Research Publication No. 16.

Municipality of Buenos Aires 1983.   Code for environmental pollution prevention (Ordinance 39025, May 31, 1983).   Ambiente y Recursos Naturales, Vol.II, No.2, 1985, p.79

National Decree 1978.   National Decree No. 2125, September 15th, 1978, Argentina.

Province of Cordoba 1985.   Law No. 7343, August 29th, 1985.

Rifkin, J. 1981.   Entropy: a new world view.   Bantam Books, New York.

Sunkel, O. 1981.   La dimension ambiental en los estilos de desarrollo de America Latina.   E/CEPAL/G 1143 (UN Economic Commission for Latin America).

Sunkel, O. and J. Leal 1984.   Economia y medio ambiente en la perspectiva del desarrollo - II. El marco conceptual: consideracion de la termodinamia (desechos y entropia).   Ambiente y Recursos Naturales, Vol.I, No.3, 1984, p.52.

Tarak, P. 1983.   Does wildlife have legal standing? The penguin case in Patagonia.   International Journal for the Study of Animal problems.   Vol.4, No.3, 1983.

Tarak, P. 1984.   Las ONG Ambientalistas se organizan para participar en la democracia Argentina. Ambiente y Recursos Naturales, Vol.I, No.4, 1984.

Tarak, P. 1985.   Hacia una Politica Argentina para los Residuos Peligrosos.   Ambiente y Recursos Naturales, Vol.II, No.2, 1985, p.42.

Tarak, P. 1986.   La politica y la legislacion ambiental: una respuesta al desafio del siglo.   In: Revista Juridica de Buenos Aires, Vol.I, A. Perrot, Buenos Aires, Argentina.

Vanosi, J.R. 1984.   El Municipio.   Cuidad Argentina, 1984, Mendoza, Argentina.

CHAPTER 13

POLLUTION HAZARDS OF AGRICULTURE

David Baldock

Institute for European Environmental Policy, 3 Endsleigh Street, London, WC1H 0DD, UK

ABSTRACT

Unlike manufacturing industry, agriculture has not been perceived as an important source of pollution. Nonetheless, there are now a substantial number of environmental hazards associated with agriculture, stemming not only from the technological revolution of the last fifty years but also from other trends such as the increasing intensity of agriculture, the tendency for production to become more specialized and the growing regional concentration of certain kinds of farming.

The nature and magnitude of the pollution hazards arising from farming vary enormously, reflecting both the differences in agricultural practice and the characteristics of the local environment. Some of the principal forms of pollution include soil erosion by water and wind, salinization arising from irrigation, increased silt load caused by land drainage, pollution of surface and groundwaters by inorganic fertilizers and livestock wastes, contamination of soils with heavy metals, multi-media pollution by pesticides and a number of other problems, such as water pollution from silage effluent and discarded pesticide containers. In most OECD countries, agriculture is responsible for a significant proportion of the nitrogen and phosphorus entering fresh waters and is a major target for those concerned in the reduction of non-point sources of pollution.

In the past, agriculture has escaped many of the regulatory pressures applied to industry and the control of non-point sources of pollution remains a relatively difficult challenge, although it may often be cost effective. In future, it seems likely that agriculture will be more susceptible to the dictates of the "Polluter Pays" principle. As well, relatively restrictive legislation is now being introduced in some European countries where groundwater pollution has become a major concern.

POLLUTION AND AGRICULTURE

The activities most commonly associated with environmental pollution include manufacturing industry, mineral extraction and processing, power generation and waste disposal. Agriculture, by comparison, tends to be viewed more as a benign activity, inherently cleaner and more sustainable than many industrial processes. Indeed there are a great many countries in which farmers traditionally have been respected as stewards of the land, husbanding their own resources and engaging in a particularly honest form of endeavour. Although agriculture has undergone profound changes in recent years, the weight of this tradition can still inhibit recognition of the full environmental impact of modern farming practices.

One of the most striking features of modern agriculture is the intensity of production, which has increased markedly in the last fifty years. Higher-yields per hectare and per animal have been achieved by employing new technologies, new varieties and new breeds, by mechanization and increasing the scale of production, by improving management techniques and deploying more capital, by reforming structures, investing in land improvement, irrigation, drainage, etc.. New technologies have introduced changes in farming systems as well as in specific practices. For example, the widespread availability of relatively cheap inorganic fertilizers made it possible for many farmers to obtain sufficient crop nutrients without keeping any livestock and thus to abandon traditional forms of mixed farming and concentrate on arable production instead.

The tendency for crop and livestock farming to be treated as separate businesses is only one example of a broader tendency towards specialization. As agriculture has become enmeshed more closely with other sectors of the economy, such as the agrochemical supply industry and the food processing and retailing sectors, it has had to become more responsive to pressures from other components of the much expanded food supply industry.   Livestock farms, as well as becoming more specialized themselves, have in many countries become concentrated in areas where feed supplies are available most cheaply.   One of the main reasons, for example, for the enormous expansion of intensive pig and poultry farms in the Netherlands in the last twenty years, has been the ready supply of low cost concentrate foodstuffs, imported in large quantities into Rotterdam.

The environmental impact of modern agriculture is not in all cases greater or more deleterious than that of more traditional forms of farming.   The ancient Greeks were familiar with the problem of soil erosion, then caused by over-grazing, deforestation and inappropriate forms of cultivation much as it still is today in parts of Greece and several other Mediterranean countries. Nonetheless, it is fair to say that the adoption of modern forms of agriculture has introduced a number of new environmental hazards and pressures, which together represent a considerable source of concern.

In considering these environmental pressures and individual pollutants of an agricultural origin it is worth bearing in mind that they arise not simply because of the use of agrochemicals and other new substances in farming, but also because of the greater intensity of production and related changes in agricultural management and the broader changes in the economy of food production.

Some of the most important environmental effects of agriculture are shown in Figure 1, prepared by the OECD secretariat.   It distinguishes a number of agricultural practices which can result in undesirable environmental changes, including the pollution of water, soil and air.   Clearly, there are many factors which determine whether or not an individual farming practice in a particular location will have a negative impact on the environment.   ome of these are complex or poorly understood.

Site specific factors are of particular importance in assessing the potential pollution hazards of many activities, such as spreading animal waste on farmland. This can make it difficult to define precisely which practices are acceptable or to draw up guidelines and rules which have general applicability.

Nonetheless, our knowledge of the pollution hazards of many agricultural activities has advanced greatly in recent years and we can identify most of the major

AGRICULTURAL PRACTICES	SOIL	GROUND WATER	SURFACE WATER	FLORA	FAUNA	OTHERS: Air, noise, landscape, agricultural products
Land development: land consolidation programmes	Inadequate management leading to soil degradation	Other water management influencing ground water table	Soil degradation, siltation, water pollution with soil particles	Loss of species		Loss of ecosystem, loss of ecological diversity. Land degradation if activity not suited to site
Irrigation, drainage	Excess salts, water logging	Loss of quality (more salts), drinking water supply affected		Drying out of natural elements, affecting river ecosystems		
Tillage	Wind erosion, water erosion					
Mechanisation: large or heavy equipment	Soil compaction, soil erosion					Combustion gases, noise
Fertilizer use — Nitrogen		Nitrate leaching affecting water				
— Phosphate	Accumulation of heavy metals (Cd)		Run-off, leaching or direct discharge leading to eutrophication	Effect on soil microflora		
— Manure, slurry	Excess: accumulation of phosphates copper (pig slurry)	Nitrate, phosphate (by use of excess slurry)		Eutophication leads: to excess algae and water-plants	to oxygen depletion affecting fish	Stench, ammonia
— Sewage sludge, compost	Accumulation of heavy metals, contaminants					Residues
Applying pesticides	Accumulation of pesticides and degradation products	Leaching of mobile pesticide residues and degradation products		Affects soil microflora; resistance of some of weed	Poisonning; resistance	Evaporation; spray drift, residues
Input of feed additives, medicines	Possible effects					Residues
Modern buildings (e.g. silos) and intensive livestock farming	See: slurry	See: slurry	See: slurry			Ammonia, offensive odours, noise, residues. Infrastructure: Aesthetic impacts

Source: OECD, 1985

Figure 1.   Selected environmental effects of agriculture.

areas of concern and turn to a growing body of research on farming and the environment. Water pollution is one of the principal areas of concern and is of particular importance for large lakes. However, there are other routes whereby agriculture can cause the pollution of lakes. Many are affected by the deposition of air-borne pesticides for example, and there may be value in taking a fairly wide view of the subject. Here, the major categories of agricultural pollution can be considered only briefly.

Soil Erosion

Soil erosion can be an important source of pollution as well as a degradation of one of the planet's primary resources. Soil erosion occurs over very large areas predominantly in drier zones, such as the Mediterranean countries, Australia and parts of the USA. There are now signs however, of increasing erosion in parts of North Western Europe as well, particularly where some of the less resilient soils are subject to intensive cultivation - in Belgium and West Germany for example. In the USA it is thought that current erosion rates are a threat to more than a third of the national cropland area and are greater than 25 tonnes per hectare per annum on about 10 per cent of cropland. Average rates of erosion vary enormously between regions and between countries. In Canada, the annual average is around 4 tonnes/hectare, in the USA about 10 tonnes and in Spain 33 tonnes (OECD 1985).

Cultivation increases erosion rates and where land is left bare of vegetation, losses are apt to be greater still. In Belgium, for example, natural erosion rates of 0.01-0.05 $kg.m^{-2}$ per annum, rise to 0.3-3.0 $kg.m^{-2}$ on cultivated land and 0.7-8.2 $kg.m^{-2}$ on bare soil (Morgan 1985). Where erosion is severe it can result in the transport of considerable quantities of silt into ditches, streams, rivers, lakes and other aquatic ecosystems. A variety of problems can arise from this, notably siltation of lakes, reservoirs and water channels, habitat damage, flooding, clogged water mains and dispersal of plant nutrients and agrochemicals present in soils. This is one route whereby pesticides, fertilizers and other pollutants reach large lakes. It has been estimated that the total annual cost of off-farm damage attributable to soil erosion in the USA is in the region of $3-13 billion at 1980 prices (Clark et al. 1985).

While soil itself can be considered a pollutant in the circumstances of wind or water erosion it is also a medium affected by pollution. "Soil pollution" is a less familiar concept than air or water pollution, but it is attracting growing attention in many European countries. A number of materials of agricultural origin can build up in the soil and if they ceach excessive levels may be regarded as pollutants. Fertilizers, animal wastes, pesticides and sewage sludge, all materials deliberately applied to farm land to enhance crop production are the main sources of pollution.

Phosphates and potassium, two of the main constituents of both artificial fertilizers and farm animal wastes, are generally bound in the top soil if not taken up by a crop. Both leach down into the groundwater only slowly in most circumstances, although leaching of phosphates can be substantial on light soils, particularly where heavy dressings of animal waste are applied regularly. There is little quantitive evidence of harmful effects arising from the accumulation of phosphates in the top soil, but high levels of potassium in the soil can inhibit plants from taking up other elements and may also affect grazing animals.

The main concern about soil pollution arising from fertilizer use arises from the presence of heavy metals in certain fertilizers.

Small quantities of chromium, cadmium, copper, lead and other heavy metals may be present in phosphate fertilizers for example. Levels of cadmium are known to be rising in some soils, and agricultural fertilizers have been established as one of the major causes of this, along with sewage sludge, industrial emissions and vehicle exhausts. Northern Europe have introduced or begun to develop new policies to regulate cadmium in the environment. In the Netherlands, for example, there is a law setting an upper limit on the amount of cadmium permissible in soil or in groundwater. There has been discussion about the need to regulate cadmium levels in phosphate fertilizers in the Federal Republic of Germany (von Moltke et al. 1986).

The spreading of animal wastes on farm land in the form of manure or slurry may also give rise to accumulating levels of heavy metals in the soil. In Europe, the best known example of this is copper, which is added to some formulations of concentrated pig feed in order to increase the animals' growth. This is subsequently excreted and then spread to agricultural land as an impurity in pig slurry. In areas where intensive pig farms are concentrated and relatively little land is available for slurry disposal, such as parts of the Netherlands, Belgium and Germany, regular heavy doses can lead to a build up of copper in the soil, with deleterious effects on the health of certain crops and in some cases resulting in pasture becoming unsuitable for the grazing of sheep, which are sensitive to copper.

A third source of heavy metals in agricultural soil is sewage sludge, a substantial proportion of which is disposed of on agricultural land in several countries, including FRG and the UK. The potential hazards of this practice include the accumulation of heavy metals and other contaminants, some of which may enter the food chain, and the spread of pathogenic organisms. There is some concern that microbiological activity in the soil may be impaired by the accumulation of heavy metals found in sludge (Brooks and McGuth 1984). In many countries there are guidelines or regulations which apply to sewage sludge application. An EEC Directive setting standards for the whole Community has been under discussion for some time.

Pesticides and other agrochemicals are another group of substances which can accumulate in the soil. Some of the more persistent forms of pesticides, such as DDT and other organochlorines, break down over many years and many of the more recently produced pesticides are less persistent. Persistent agrochemicals remain in use, however, and there are some where the regular spraying of vines and hops with cupriferous fungicides has resulted in copper accumulating in the soil to the extent that serious changes in soil biology have occurred. The effects of pesticides on soil life in the longer term are not entirely understood and while there is little evidence of the reproduction of microorganisms being impaired by pesticide residues, soil fauna, including earthworms, are more vulnerable and adverse changes can occur. Even where pesticides break down rapidly in the soil there remains considerable concern about their effects when leached out of the soil into the aquatic environment, an issue returned to later.

Air Pollution

Although not always perceived as a source of air pollution, agricultural activity can give rise to a number of different pollutants, including pesticide spray drift, ammonia arising from livestock wastes, smoke and smuts from the burning of straw, dust and unpleasant odours, especially from intensive livestock units and farm slurry spreading.

Some of these air-borne substances are a source of serious concern. Ammonia in the Netherlands is one example, while others, such as odours from slurry storages are very local in their impact and often are classified as "nuisances" rather than pollution. Local conditions often determine when a nuisance comes to be regarded more seriously. In the UK, for example, there has been growing opposition to the burning of surplus straw in the field after harvest because it has resulted, among other things, in pollution, damage to property from uncontrolled fires and an unnecessary load on local fire services. However, following a number of cases where smoke has enveloped main roads and even caused a fatal accident, the hazards of straw burning have been perceived as more serious, with stronger penalties being introduced and growing pressure for a total ban.

Where pesticides are carried on the air beyond their immediate targets, they can cause damage to neighbouring crops, livestock and wildlife, enter water bodies and also create a hazard for human health. "Spray drift", or the transport of small droplets of agrochemical spray away from the target crop on air currents is one of the ways in which this occurs, and is likely to be the most severe where aircraft or helicopters are used for spraying. However, certain agrochemicals may drift some days after they have been applied. Ester formulation herbicides are particularly known to volatilize under certain conditions and subsequently drift on the wind in the form of a very fine vapour. Consequently, agrochemicals of this kind tend to spread into the wider environment rather than being confined to specific cropping areas.

With "acid rain", a central environmental concern in many European countries, attention has been drawn to one of the less familiar contributing factors, ammonia emissions originating from livestock wastes. In the Netherlands, where livestock production is particularly intensive and waste management has become a severe problem, there is concern about ammonia emissions contributing to the oxidation of SO_2 in the atmosphere to form sulphuric acid as well as about the effects of ammonium sulphate deposition. This is one of the major reasons for the recent introduction of regulations to control intensive livestock farms in the Netherlands.

Water Pollution

Perhaps the most important category of pollution associated with agriculture is water pollution, covering both ground and surface waters. As already illustrated, soil and water are interconnected parts of a larger system, the workings of which may be affected by agriculture in many different ways. Water quality is not always the prime concern. Farming practices also influence the quantity of water available within an ecosystem and while this paper will not attempt to deal with the ecological impact of land drainage and irrigation schemes, these can be of great significance. As the German Council of Environmental Advisors (Der Rat von Sachverstaendigen fuer Umweltfragen) has noted in an authoritative recent report on agriculture and the environment:

> ..."Every use of land for agricultural purposes has specific effects on the quantity, distribution and composition of water. In addition to the direct effects of irrigation and drainage measures on a water balance, the type and timing of cultivation and manuring measuces, as well as the type and duration of the vegetation cover can also influence the quantiy and quality of a water"... (Council of Environmental Advisors 1985).

Some of the most widespread agricultural pollutants include nitrate ion, phosphate ion, agrochemicals, soil and a variety of farm wastes containing organic matter, such as slurry and silage. However, a more comprehensive list would include potassium ion, heavy metals, oil, used pesticide containers, veterinary products and other materials utilized in modern farming.

Agriculture is one of the principal sources of nitrogen and phosphorus in many of the surface waters including lakes, large and small, where eutrophication is taking place. Whereas both nutrients are required for eutrophication to occur, it is now widely accepted that phosphorus is usually the factor that determines the development of the process and that even where this is is not the case and some other nutrient, usually nitrogen, has the decisive role, it may still be possible to adopt appropriate control policies such that phosphorus can be made to play the role of the limiting factor (OECD 1982).

Sources of phosphorus entering lakes are thus of considerable interest. Conventionally these are divided into "Point- and Non-point" or "Diffuse" sources. Diffuse sources are less easily measured and may be more difficult to regulate than point sources, such as sewage works and industrial effluent pipes. However, they are of great importance for many lakes. Table 1 gives a breakdown of the diffuse sources of phosphorus load entering lakes via tributaries, which includes several kinds of agricultural effluents.

Studies in the Great Lakes region have suggested that cropland is the major source of diffuse phosphorus loads and that land-use activities, primarily

TABLE 1   DIFFUSE SOURCES OF PHOSPHORUS INPUTS
          CONTRIBUTING TO THE PHOSPHORUS LOAD OF LAKES
          VIA TRIBUTARIES[1]

a)   Effluents from non-sewered, populated areas

     - septic tanks;
     - domestic wastes;
     - soil erosion.

b)   Effluents from cultivated land

     - soil erosion;
     - fertilizer losses;
     - domestic animal excrements;
     - organic plant wastes.

c)   Effluents from non-cultivated land

     - soil erosion;
     - organic plant wastes;
     - wild animal excrements.

d)   Spring and natural waters

e)   Reserves in bodies of water

     - sediments;
     - fauna and flora;
     - groundwater.

f)   Atmosphere
     - wet precipitation;
     - dry precipitation.

[1]OECD, Eutrophication of Waters: Monitoring.
Assessment and Control, Paris, 1982

agriculture and forestry, contribute between a third
and a half of the total phosphorus entering the Great
Lakes (Clark et al. 1985).   Italian researchers, using
a   theoretical   approach,   have   calculated   that
agriculture (excluding farm animals) contributed an
average of 21 per cent of total phosphorus loadings in
the EEC countries in 1982.   Ireland is at the top end
of the range, with an estimate of 34 per cent and the
Netherlands at the lower end, with a figure of 15 per
cent.   Farm were considered separately and contributed
a further 18 per cent on average, ranging from 36 per
cent in Ireland to 13 per cent in Italy (Vighi and
Chiaudini 1985).   In general terms, the OECD suggest,

agricultural activities account for about 30 per cent of phosphorus loadings in many rural areas and also for 70-85 per cent of the total nitrogen loading (OECD 1982).

While efforts to control phosphorus loadings have concentrated mainly on point sources, there are situations where the regulation of diffuse sources may be more cost-effective. Table 2 gives some cost estimates applicable to Lake Erie in the late 1970s and suggests that a phosphorus reduction of about 10 per cent could be expected simply by improving agricultural management, at no cost, while more extensive changes in practice may still be cheaper than reducing sewage effluent total phosphorus concentrations below 0.5 mg.L^{-1}.

In many western European countries the form of agricultural pollution attracting greatest interest is nitrate contamination of drinking water. Nitrate concentrations in groundwater are rising over extensive areas and a similar growth is occurring in some surface waters used for drinking purposes. The main health hazards associated with nitrate consumption are methaemolglobinaemia and, more controversially, gastric cancer. For many years there has been a World Health Organization recommended limit on nitrate concentrations of 45 mg.L^{-1}, but in July 1985 an EEC Directive on drinking water quality (No 80/778) came into force, setting a maximum admissible concentration of 50 mg.L^{-1}. Surveys suggested that about eight per cent of Danish sources of potable water did not always meet this standard (National Agency of Environmental Protection 1984) and the maximum is also exceeded elsewhere in Europe, in Germany, France and the UK for example.

In many rural areas, nitrate concentrations in water seem to be rising steadily and in Denmark, studies suggest that a mean value in drinking water of around 4 mg.L^{-1} prior to 1960 has climbed to over 13 mg.L^{-1} today. In a fairly recent study of 75 groundwater sites by the Department of the Environment in the UK, it was found that about 65 per cent showed rising trends and a further 30 per cent were categorized as apparently static, although subject to substantial variations (STACWQ 1984). Agriculture is primarily responsible for nitrate pollution, but factors are involved in determining the rate at which nitrates leach into groundwater and reach surface waters. As well, there is considerable uncertaintv about the extent to which deep aquifers will be affected by the marked growth in synthetic fertilizer use which has taken place over the last two decades. In some areas,

TABLE 2   ESTIMATED COSTS OF PHOSPHORUS REDUCTION
          ALTERNATIVES AVAILABLE TO LAKE ERIE (1979)

Remedial measure options	Estimated annual incremented unit costs - $/kg phosphorous reduction
Urban point sources:	
a) 1.0 mg/l to 0.5 mg/l	8.0
b) 0.5 mg/l to 0.3 mg/l	95.5
Rural nonpoint sources:	
Level 1 Sound management on all agricultural lands, avoiding excess fertilisation, reducing soil erosion (10% phosphorous reduction)	
Level 2 Level 1 measures, plus bufferstrips, strip cropping, improved municipal drainage practices, etc., depending on region (25% reduction in phosphorous losses on soils requiring treatment)	64.3
Level 3 Level 2 measures at greater intensity of effort (to achieve 40% reduction in phosphorous losses on soils needing treatment)	174.0
Urban nonpoint sources:	
Level 1 Programme of pollutant reduction at source	82.0
Level 2 Level 1 measures, plus detention/ sedimentation	156.9

Source: OECD, Eutrophication of Waters: Monitoring, Assessment
and Control, 1982

in the Netherlands for instance, livestock wastes are
thought to be the main contributors to nitrate
pollution especially when applied in excessive
quantities in the late autumn and winter.

Fertilizer use in many parts of Europe tends to be
higher than in North America and with an annual
application rate of 180 kg per hectare, the OECD
suggest that annual leaching losses may be about 30-45

kg per hectare, or between 17 and 25 per cent of the amount supplied. In seeking to maintain acceptable drinking andards, European countries have examined options, including better advice for farmers, blending and denitrification of water supplies, the control of fertilizer use within water protection zones and the imposition of a tax on synthetic nitrogen fertilizers. Sweden has recently decided to levy a tax on nitrogen fertilizers and the Netherlands has introduced mandatory upper limits on animal waste applications per hectare per annum with the phosphate content being treated as the limiting factor.

Apart from the nitrogen and phosphorus content of farm animal wastes, their typically high biological and chemical oxygen demand can give rise to considerable short-term pollution problems, especially in small rural water courses near slurry storages or bordering fields and yards where run-off occurs. Slurry usually has a high ammonia content and is harmful to fish and thousands fish can be killed in incidents arising from bad agricultural management, when slurry storages overflow into neighbouring rivers for example. Several other waste products also have a high oxygen demand, including dirty yard water, waste milk and silage effluent which is now one of the main surface water pollutants in livestock rearing areas of Britain.

In a rather different category are agrochemicals, including pesticides, herbicides, fungicides, agaricides and other products designed to protect crops and livestock. Many factors play a part in determining whether these products and their metabolites reach surface or groundwaters and in what form and quantity. Concern is usually focussed on the most persistent agrochemicals, such as DDT, those which are most mobile in the soil and so likely to leach and those which are most toxic. The leaching of pesticides into groundwater is occurring in several countries, although as would be expected, the concentrations are low. As this trend has become more apparent, and lower concentrations have become detectable, there has been a growth of interest in monitoring and research. Indeed, the hazards of groundwater contamination may have been significantly underestimated in the past. Herbicide residues in drinking water in parts of the UK are above the maximum acceptable concentration of 0.5 ug/L set in EEC Directive 80/788, although they are not being treated as a hazard at present and the government appears to think that this standard may be unnecessarily severe (Department of the Environment 1986).

Severe pesticide pollution incidents can affect water supplies as a result of bad agricultural practice or accidents.   Water bodies may be accidentally sprayed with toxic products.   Some fungicides will kill fish for example, tanks may leak, or be washed out into a watercourse, highly toxic sheep dip chemicals may be disposed of incorrectly and old pesticide containers may be allowed to come into contact with water sources. Particular care has to be exercised with aquatic herbicides, many of which will kill fish and other aquatic life if improperly used.

In general, the trend is to try to replace highly persistent pesticides with more recently developed and less persistent alternatives.   Many governments operate regulatory regimes whereby pesticides must be tested and cleared for use.   Application techniques tend to be less closely regulated and relatively few countries have mandatory training courses for spray operators.

Wider Environmental Effects

The impact of agricultural practice on the environment extends substantially beyond the categories of pollution considered briefly here.   In OECD countries as a whole, agriculture occupies on average about 40 per cent of the land area.   In several countries the proportion is over 70 per cent.   Over vast areas the development of farming has had a profound effect on ecosystems and landscapes.   Some of the most dramatic effects of agricultural development are evident in newly irrigated areas, in drained marshland, in those parts of rural Europe where land consolidation is still taking place and new landscapes are designed on drawing boards.   However, a host of smaller scale changes are occurring at the same time as fields become larger and old boundaries are removed, as areas of wildlife habitat are converted into farmland, old fields and terraces are abandoned and orchards are grubbed to be replaced with arable fields.

The effects of agriculture on wildlife habitat and landscape are of particular concern in many European countries, especially those where intensive farming methods are most widely used.   In the United Kingdom, the effects of agriculture on wildlife habitats and the landscape have generated a passionate debate and put the issue high on the environmental agenda, whereas concern about farm pollution is markedly less intense. Wildlife is affected by agricultural activities in a large number of ways.   Direct effects, such as pesticide poisoning, are usually less important than longer term changes in vegetation cover, ecosystems and soil biology.

Another aspect of agriculture which has not been discussed here is the presence of residues and impurities in crops and livestock products. The presence of pesticide residues in food is one of the best known of these contaminants. There are also several others causing some degree of concern, such as the presence of artificially implanted hormones in beef and veal, elevated nitrate levels in some intensively fertilized vegetables and the residues of pharmaceutical products in some meat and milk.

Policy Responses

In 1979 the Royal Commission on Environmental Pollution completed a detailed and widely respected study of agriculture and pollution in the United Kingdom. In their conclusions they observed that with regard to pollution it was frequently assumed that:

> ..."agriculture is a special and privileged case; that the necessity of food supply takes precedence over environmental requirements and that the observance of good agricultural practices will ensure that any adverse environmental effects are reduced to a practicable and acceptable minimum"...

They went on to comment:

> ..."This view may well be justified for agriculture as it was traditionally practised up to the end of the Second World War. Since that time, as we have described, changes have taken place that greatly increase the impact of agriculture on the environment and which suggest the need for reassessment"...(Royal Commission on Environmental Pollution 1979).

This reassessment is now proceeding in several different countries and on a range of different fronts. While agricultural interests have generally been successful in resisting the kind of regulations now commonly applied to manufacturing industry, it has become more difficult to argue that pollution hazards from agriculture are too trivial to merit special attention. Understanding of the environmental consequences of different agricultural practices has increased considerably and this had led to more specialized research, improvements in official advice, and in several countries greater attention to codes of good agricultural practice.

While advice, exhortation, codes of good practice and voluntary agreements with farmers still play a major part in government strategies to limit agricultural pollution, there is also a growing body of legislation setting mandatory standards, imposing planning procedures, banning certain substances and practices and encouraging "environmentally friendly" farming practices. Pesticide legislation exists in many OECD countries and pesticide residues are the subject of legislation covering all EEC countries. Attempts in the United States to control soil erosion were greatly strengthened by some of the provisions of the 1985 Farm Bill, which also introduced new provisions covering "sod busting" and "swamp busting". Tightening drinking water quality standards in the EEC has forced governments to consider ways of limiting nitrate leaching and run-off from farmland and has helped to spark a debate on whether farmers should be subject to the "Polluter Pays" principle.

Regulation of farming practices has often been regarded as impractical in the past, but recent Dutch regulations restricting animal waste spreading on farm land to specified maxima, suggest that legislators feel less inhibited by practical enforcement difficulties than in the past. Maximum application rates for slurry have also been adopted as part of a broad package of new measures applying to farmers in Denmark known as the "Handlingsplan". The law is highly specific about the design and siting requirements for animal housing, livestock waste storage facilities, silage pits, etc..

While other countries with less intensive agriculture and greater reticence about environmental legislation are likely to tread more cautiously, there is little doubt about the direction of change. In 1985, six years after the Royal Commission reported, the Council of Environmental Advisors in Germany began their overview of their study of the environmental problems of agriculture with the words:

> ..."The contamination of intensive farming requires that drastic environmental and agricultural policy measures are introduced with the object of reversing the pollution trend and restoring the biotope function of the agrarian landscape"...(Council of Environmental Advisors 1985).

In at least one country, the concept of reversing agricultural pollution seems to have taken hold.

REFERENCES

Brooks, P.C. and McGuth, S.P. 1984.   Effects of metal toxicity on the size of the soil microbial biomass. Journal of Soil Science, Vol 35, pp 341-346

Clark, E.H. Haverkamp, J.A. and Chapman, W. 1985. Eroding soils: the off-farm impact.   The Conservation Foundation, Washington, DC.

Council of Environmental Advisors 1985 March.   Summary of the report on environmental problems of agriculture. Geschaftsstelle des Rates von Sachverstaendingen fuer Umweltfragen, Wiesbaden, FRG

Department of the Environment 1986.   Long Term Water Research Requirements Committee.   Water research in the longer term.   Department of the Environment, London.

Morgan, R.P.C. 1985.   Soil erosion measurement and soil conservation research in cultivated areas of the UK. The Geographical Journal, Vol 151, No 1, pp 11-20.

von Moltke, K., Bauman, P. and Irwin,F. 1986.   The regulation of existing chemicals in the European Community: Possibilities for the development of a community strategy for the control of cadmium.   Report for the EC Commission, IEEP, Bonn.

National Agency of Environmental Protection 1985.   The NPO Report.   NAEP, Copenhagen, Denmark.

OECD 1982.   Eutrophication of Waters: Monitoring, Assessment and Control. OECD, Paris.

OECD 1985.   The State of the Environment 1985. OECD, Paris

Royal Commission on Environmental Pollution 1979. Seventh Report: Agriculture and Pollution.   HMSO, London.

STAWCQ (Standing Technical Advisory Committee on Water Quality) 1984.   Fourth Biennial Report February 1981- March 1983. Department of the Environment, London.

154   David Baldock

Vighi, M.e and Chiaudini, G. 1985.   The impact of
agricultural loads on eutrophication in the EEC surface
waters.   In: <u>Environment and Chemicals in Agriculture</u>.
ed F.P.W. Winteringham, pp 71-83, London, Elsevier
Applied Science Publishers.

CHAPTER 14

STRATEGIES FOR CONTROL OF NON-POINT SOURCE POLLUTION

Norman A. Berg

American Farmland Trust, 1717 Massachusettes Avenue,
N.W., Washington, D.C.

ABSTRACT

In 1972, the U.S. Congress, in amending the U.S. Water
Pollution Control Act (P.L. 92-500) set a national goal
to eliminate all water pollutant discharges by 1983.
The nation's waters are not yet pristine and point-
source water pollution controls alone are proving
insufficient to meet the objectives of the Clean Water
Act. The final report of PLUARG, Environmental
Management Strategy for the Great Lakes System, to IJC
in 1978, is still valid. It resulted from an extensive
six year study of the impact on water quality from land
used for a variety of activities, including
agricultural, forestry, transportation, urban
development and waste disposal. Pollution from non-
point sources is best characterized by the large
numbers, the wide variety, the intermittent nature of
inputs, the seemingly insignificant nature of the
individual contributions, the damaging effect of the
cumulative impact, and a wide variety of economic,
social and institutional interactions playing a role.
Key elements of the recommended strategy included
regional prioritization of problem areas, expansion of
soil erosion and animal waste control programs,
comprehensive management strategy, and incorporating
the load reduction schedules for phosphorus in the
Great Lakes Water Quality Agreement between Canada and
the U.S.A. After nearly a decade of implementation of
the four major components of the "Strategy"

(information, education and technical assistance,
planning, fiscal arrangements and regulation, current
problems) are analyzed. The recommendations in 1978 by
PLUARG provided a solid foundation for accelerating
non-point source programs. Flexible management systems
and control measures capable of incremental adjustments
in response to a changing environment will be required.

INTRODUCTION

On August 18, 1978, the Chairmen of the International
Join Commission (IJC) Robert J. Sugarman (United States
Section) and Maxwell Cohen (Canadian Section), signed a
letter. It was addressed to Dr. Murray G. Johnson and
the author as co-chairmen of the International
Reference Group on Great Lakes Pollution from Land Use
Activities (PLUARG). That letter, quoted in part,
stated:

> ..."Thank you very much for your excellent
> presentation in Windsor and for your years of
> intensive research and discussion by all
> members of the Group and its Committees.
> Your explanation of the conduct and the
> results of this very exhaustive study, and
> your presentation of its findings, were much
> appreciated. Your study will be regarded as
> a landmark in the consideration of non-point
> sources of water pollution."...

The 1972 Canada-United States Agreement on Great Lakes
Water Quality (GLWQA) had directed the IJC to conduct a
study of pollution of the boundary waters of the Great
Lakes System from agricultural, forest and other land
uses.   IJC had established PLUARG as a bi-national
group of scientists to develop a work plan and a
schedule to do the needed research and analysis to
answer the following three questions as set forth in
the terms of reference of the 1972 Agreement:

> 1.Are the boundary waters of the Great Lakes
> System being polluted by land drainage
> (including ground and surface runoff and
> sediments) from agriculture, forestry, urban
> and industrial land development, recreational
> and parkland development, utility and
> transportation systems and natural sources?
>
> 2.If the answer to the forgoing question is
> in the affirmative, to what extent, by what

causes, and in what localities, is the pollution taking place?

3. If the group should find that pollution, of the character referred to, is taking place, what remedial measure would, in its judgement, be most practicable and what would be the cost thereof?

The scope of this inquiry was broader than previous Great Lakes studies conducted under the sponsorship of the Commission in that the entire land area, as well as the water, in the Basin was studied. The Basin totals 755,200 km^2 in area, with 538,900 km^2 of land and 216,300 km^2 of water surface area. The Great Lakes contain approximately 20 percent of the world's fresh surface water supply. Eutrophication, due to elevated nutrient inputs, particularly in the two lower lakes (Erie and Ontario) and the increasing contamination of these water bodies by toxic substances, had been identified as the major pollution problems in the Basin. Increasing population, advancing technological innovation and intensification of the land and water use in the Basin had resulted in a continuing degradation of the Great Lakes.

It had also become apparent that while the lakes themselves were a focal point of concern, they are part of a complex system. The basic thrust of the PLUARG work recognized that the ultimate effects of any pollutant involves a complex interaction of land, water and organisms. Therefore, a study of pollutant effects, regardless of the source, which did not recognize these chemical, physical and biological components, could produce a partial or even misleading understanding of such effects. This perspective of the "Great Lakes Ecosystem" was also advocated by the Research Advisory Board in their role as principal scientific advisor to the International Joint Commission on Great Lakes Water quality (Great Lakes Research Advisory Board - Annual Report to the International Joint Commission [Windsor, Ontario] July 1977). Although the Great Lakes are an interconnected system, each basin is unique in terms of its limnology, the socio-economic characteristics of its communities, the type and degree of pollution and the control measures required to solve the pollution problems. Non-point source pollutants are not derived uniformly from whole water sheds or even sub-basins. Problem areas could well represent only a small proportion of a drainage basin area. As a result PLUARG developed criteria for the identification of potential contributing areas and within these, the most hydrologically active areas (HAAs), which are most likely to produce water pollution from land use activities.

In the study a substance was considered a pollutant on the basis of two criteria:

    1.demonstrable adverse effects on water quality or biota in either the nearshore zone or offshore waters of the Great Lakes, and

    2.the substance had derived largely from diffuse sources.

Direct atmospheric and in-lake sediment sources were considered, although in the strictest sense, these inputs did not constitute land drainage sources. Many of the substances identified as Great Lakes pollutants are required by many aquatic organisms for growth and reproduction. It is in excessive quantities, relative to these needs, that they represent a real or potential hazard to the Great Lakes ecosystem, and are defined as pollutants.

It is also important to recognize:

    1.the long term nature of the solutions to most problems of pollution from land use activities,

    2.their ramifications through most sectors of society,

    3.the involvement of many agencies in the implementation of these solutions, and

    4.their public consequences in such policy areas as food production and public health.

Population growth and location, industrial development and technological innovation all have impacts on the loadings to the lakes from both point and non-point control measures and the ability to control some of these sources. As populations grow and industrial development continues given current technology, pollutant inputs will undoubtedly continue to increase from both point and non-point sources. Therefore, the strategy for non-point source reduction is increasingly important. This strategy must also transcend jurisdictional and political boundaries. Questions of equity were also taken into account. A formula was developed for the reasonable allocation of responsibility between governments, institutions and individuals. PLUARG recognized that the management of non-point sources would require a departure from the traditional point source control approach.

HOW DID PLUARG APPROACH THE STUDY?

The international reference group was composed of nine Canadian and nine United States government officials. Reasons for the reference by IJC and purposes of the study were:

> 1.to determine and evaluate the causes, extent and locality of pollution from land use activities,

> 2.to gain an understanding of the relative importance of various land uses in terms of their diffuse pollutant loads to the Great Lakes,

> 3.to examine the effects of the diffuse pollutant loads on Great Lakes water quality, and

> 4.to determine the most practicable remedial measures for decreasing the diffuse pollutant loads to an acceptable level and the estimated cost of these measures.

Detailed plans for this study were developed in early 1973 and approved by the IJC. Assignments were made to both Canadian and United States agencies and qualified individuals to commence research on specific tasks and programs within the PLUARG framework plan. These detailed plans were subsequently updated in 1976. During the study, supporting papers and reports were developed in six broad categories (Annotated Bibliography of Reports, IJC 1977) as follows:

> -State-of-the-Art Assessments,
> -Land Use Inventories,
> -Pilot Watershed Studies,
> -Assessment of Great Lakes Water Quality
> Improvement
> -Annual Reports, and
> -Special Reports.

Major jurisdictions involved in the Great Lakes Basin are the federal governments of Canada and the United States of America, the province of Ontario and the states of Minnesota, Wisconsin, Illinois, Indiana, Michigan, Ohio, Pennsylvania and New York. There were

207 U.S. counties and 39 Canadian counties, districts
and regional municipalities in the Basin.

The major land uses inventoried for the study were as
follows:

GREAT LAKES BASIN	URBAN LAND	RURAL LAND	TOTAL
	(1,000 hectares)		
UNITED STATES	1,356.4	29,112.1	30,468.5
CANADA	354.4	23,067.0	23,421.4
TOTAL	1,710.8	52,179.1	53,889.9

Sixty-one percent of the basin consists of
forested/woodland. Twenty-four percent is agricultural
land, including cropland and pasture. Urban land,
including residential, commercial and industrial areas,
makes up three percent of the Basin. The remaining 12
percent consists of recreational lands, wetlands,
transportation corridors, waste disposal sites,
extractive industries and idle lands.

Additional background information on characteristics of
the total drainage area was collected on geology,
soils, mineral resources, climate, hydrology,
vegetation, wildlife, waste disposal operations, high
density non-sewered residential area, recreation lands,
economic and demographic data and use of pesticides,
commercial fertilizers, agricultural manures and
highway salts. In addition, trends in land use
patterns and practices in 1980 (and to 2020, where
appropriate) were made. Information from this
inventory was also used to gain a better understanding
of the combination of factors that affect pollution
from non-point sources.

In order to evaluate the extent, causes and localities
of pollution from land drainage, several areas in the
Basin were selected for detailed studies. These pilot
watersheds were designated to represent the full range
of land use activities and to permit the extrapolation
of results to the full Basin.

These pilot watersheds, as indicated in Figure 1,
included the Genesee River, selected to study the
effect of diverse land uses on water quality. The
watershed of the Genesee contains significant amounts
of agricultural, forested and urban land. The
investigation focused on identifying the combination of
factors that affect the movement and transport of

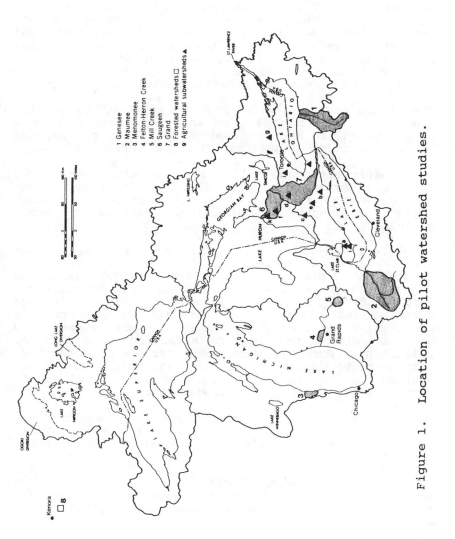

Figure 1. Location of pilot watershed studies.

phosphorus, suspended solids and chloride from the watershed to the Great Lakes.

The Maumee River watershed investigations dealt with the effect of agricultural practices on water quality. The Maumee River, the largest tributary to Lake Erie, has more than ninety percent of its land area in agricultural use. Research in this watershed was concentrated on the generation of sediment and nutrients from intensely cultivated cropland under the prevailing management practices at different times of the year and a comparison of these losses with the yields of these same materials downstream.

The Menomonee River watershed was selected to examine the impact of urban land on water quality. It is a highly urbanized area and contains land in the process of conversion from rural to urban land uses.

The Felton-Herron and Mill Creek sub-watersheds served as a focus for studying the effects of intensive use of pesticides and fertilizers in an orchard area under different practices, including wastewater spray irrigation.

The Saugeen River watershed included large areas that are in forests. Phosphorus, nitrogen and chloride inputs were studied. The Grand River Watershed represented a combination of urban and agricultural land uses. It is the largest Canadian watershed draining into Lake Erie. Studies focused on the progressive pollution from the headwaters to the mouth and on land-related factors affecting this type of pollution. Forested watersheds (12 small watersheds within the headwaters of the English and Winnipeg River systems near Kenora, Ontario), were studied to determine the effect of clear-cutting.

Ontario agricultural watersheds (11 small sub-watersheds), were selected for special study. These represented major agricultural regions in southern Ontario. They had a wide range of land characteristics, with soils varying from low to high clay content. Several different crop covers were examined along with livestock, surface hydrology and ground water movement for the impact on pollution from agriculture.

These pilot watersheds were established as soon as possible since PLUARG felt need for detailed site specific evidence that non-point source pollution was a problem impacting water quality. The studies indicated that land use is not the only land-associated factor influencing Great Lakes water quality. PLUARG identified additional factors contributing to the variances in unit area loads observed for single dominant land uses.

The most important factors that determined the magnitude of pollution from land use activities were the physical, chemical and hydrological characteristics of the land, land use intensity and materials usage. Weather was also important.

An understanding of these factors and the way they influence non-point source pollution is essential. The evaluation of these factors led to the identification of those portions of a watershed which are more hydrologically active than other areas of the same watershed. This is important in devising the strategy for reducing pollution from non-point sources that originate from agriculture. A hydrologically active area is an area within a watershed that produces significant amounts of runoff, even during relatively minor rainfall and snowmelt events. Areas with predominantly flat slopes and poorly drained soils and which are located near enough to a water body that runoff waters are delivered efficiently, can be HAA's, if soil moisture content is at a level that seriously reduces the infiltration of additional water. Under these conditions less active areas may become more active. The agricultural pilot watersheds presented examples in which 15-20 percent of the land surface contributed up to 90 percent of the total sediment load from the watershed. Because soil moisture content and land use management often vary with the season, the size of the HAA's and the importance of land management can vary with the time of the year.

Land characteristics, including soil type, surficial geology, geomorphology and soil chemistry were inventoried. While most sites have certain unique characteristics, generalizations concerning the importance of basic land "forms" have emerged for the basin. The most important is soil type, indicated by differences in soil texture or particle size. Runoff is greater from fine-grained, low permeability soils.

The coarser grained sandy soils have higher infiltration rates than clay soils. Pollutants tend to associate with the clay sized soil particles, since these particles are suspended readily by rainfall impact and runoff and settle out only in very slow flowing water. Further evidence of the influence of soil type was observed in the better water quality, with respect to sediments and phosphorus, in areas with sandy soils (e.g. the upper Lake Michigan basin). These soils have coarse sized particles and higher infiltration rates than the areas with clay soils (e.g. Lake Erie basin) with similar land use, but lower water quality. There was a high potential for clay sized particles being transported to the lakes.

Therefore, PLUARG, based on the pilot watershed studies and other research, emphasized the importance of physiographic characteristics, such as slope and drainage density, when explaining problems associated with specific sites. For example, assuming a constant clay content, a clay soil on a steep slope represents a greater pollution potential than a sandy soil on a steep slope. Clay soils on flat land represent a lesser hazard, for the potential for the movement of pollutants to receiving waters increases with greater drainage density. Surface soil and vegetation affect the amount of precipitation infiltrating, along with the intensity with which the land is used. How the land is farmed is also a major factor affecting potential pollutant contributions from land areas.

Any land practice that exposes soil to the erosive forces of rainfall, runoff or wind increases the pollution potential. In general, the greater the canopy and ground cover protection, the lower the erosion potential. Of the cultivated lands in the Basin, widely spaced row crops contribute the greatest quantities of sediments and associated pollutants. Phosphorus loads also originate on feedlots, barnyards, manure storage areas and on farm land receiving winter-spread manure. In a number of the pilot watersheds that were agricultural, these sources contribute about 20 percent of total agricultural phosphorus loadings. However, the range of values are wide. Livestock density, proximity to water bodies, adequacy of the animal waste management systems all play a role.

At the time of the PLUARG study, in the 1970s, many farmers practiced fall plowing for a variety of reasons. Many soils in the Great lakes basin are wet, and more difficult to till in the early spring. This caused many farmers to plow during the fall. This, farmers said, generally increased yields. It also exposed the soil for a longer period, increasing erosion potential. The timing and type of tillage practices do affect the amount of soil exposed to possible erosion. The farmers also were following a long tradition with fall plowing. The more recent move to more "conservation tillage" may counter that tradition. It will also reduce the potential for erosion. Larger, more continuous row cropping (mono-culture) systems lead to higher sediment loads. Farming close to streams reduces vegetative buffers and increases the risk for soil transport to the watercourse or drainage system.   Materials applied to the land, including commercial fertilizers, pesticides and manure can influence the quality of runoff.

HOW DID PLUARG ANSWER THE THREE IJC QUESTIONS?

PLUARG did find that the Great Lakes are being polluted from land use activities and subsequent runoff. Sediments, phosphorus, some industrial organic compounds, some previously-used pesticides and potentially some heavy metals, principally from areas of intensive agricultural and urban use, were impacting water quality.   Some of the problems are residual, e.g. banned or restricted pesticides which though declining, remain in the ecosystem.   The more detailed answer is reflected in Table 1, showing Great Lakes water quality pollutants.

The IJC, two years later in 1980, for the most part accepted PLUARG's findings.   However, IJC expressed reservations on certain phosphorus control issues, and placed additional emphasis on the growing problem of controlling toxic substances.   PLUARG had stated that toxic substances, such as PCBs have been found to gain access to the Great Lakes System from diffuse sources, especially from atmospheric deposition.   PLUARG also found that phosphorus loads from land drainage and atmospheric deposition contribute to both surface offshore and nearshore water quality problems related to eutrophication.   Point source control programs alone would be sufficient to meet target loads only in Lakes Superior and Michigan.   While in many cases it is difficult to ascribe pollution (e.g. violation of a specific existing or proposed water quality objective) to any particular land use, it is important to note that it is the cumulative effect of a variety of land use activities that ultimately contributes to pollution of the Great Lakes from non-point sources.

PLUARG did find that the lakes most affected by phosphorus and toxic substances are Erie and Ontario. Local problems associated with phosphorus and sediment are in such areas as Green Bay, Saginaw Bay, southern Georgian Bay, Lake St. Clair, the Bay of Quinte, and the south shore red clay area of Lake Superior. Intensive agricultural operations have been identified as the major non-point source contributor of phosphorus.   Erosion from crop production on fine textured soils and from urbanizing areas, where large scale development have removed the natural ground cover, were found to be the main sources of sediment.

TABLE 1   GREAT LAKES WATER QUALITY POLLUTANTS

I. PARAMETERS FOR WHICH A GREAT LAKES WATER QUALITY PROBLEM HAS BEEN IDENTIFIED

POLLUTANT	PROBLEM		SOURCES				REMARKS
	Lakewide	Nearshore or Localized	DIFFUSE Land Runoff	DIFFUSE Atmosphere	DIFFUSE In-Lake Sediments	POINT	
Phosphorus[i]	Yes	Yes	Yes	Yes	Yes[a]	Yes	[a] percentage unknown; not considered significant over annual cycle
Sediment[b,i]	No	Yes	Yes[c]	Negligible	Under some Conditions	Negligible	[b] may contribute to problems other than water quality (e.g., harbor dredging) [c] including streambank erosion
Bacteria of Public Health Concern	No	Yes	Minor[d]	No	No	Yes	[d] land runoff is a potential, but minor source; combined sewer overflows generally more significant
PCBs[i]	Yes	Yes	Yes	Yes	Yes	Yes	
Pesticides[i] (past)	Yes[e]	Yes[e]	Yes	Yes	Yes	No	[e] some residual problems exist from past practices
Industrial Organics	Yes	Yes	Yes	Yes	Yes	Yes	
Mercury[i]	Yes	Yes	Minor	Yes	Yes	Yes	
Lead[i]	Potential[f]	Potential[f]	Yes	Yes	Yes	Yes	[f] possible methylation to toxic form

II. PARAMETERS FOR WHICH NO GREAT LAKES WATER QUALITY PROBLEM HAS BEEN IDENTIFIED, BUT WHICH MAY BE A PROBLEM IN INLAND SURFACE WATERS OR GROUNDWATERS

POLLUTANT	PROBLEM		SOURCES				REMARKS
	Lakewide	Nearshore or Localized	DIFFUSE Land Runoff	DIFFUSE Atmosphere	DIFFUSE In-Lake Sediments	POINT	
Nitrogen	No	No[g]	Yes	Yes	Minor	Yes	[g] some inland groundwater problems
Chloride	No	No[h]	Yes	Negligible	No	Yes	[h] some local problems exist in nearshore areas due to point source
Pesticides[i] (Present)	No	No	Yes	No	No	Yes	[i] new pesticides have been found in the environment; continued monitoring is required
Other Heavy Metals	Potential[f]	Potential[f]	Yes	Yes	Yes	Yes	
Asbestos[j]	No	Yes	No	?	Yes	Yes	[j] see Upper Lakes Reference Group Report
Viruses[k]	← No Data Available →						[k] better detection methods needed
Acid Precipitation	No	No[m]	No	Yes	No	No	[m] a potential problem for smaller, soft water, inland lakes

[i] Sediment per se causes local problems; phosphorus and other sediment-associated contaminants have lakewide dispersion.

The most important land related factors affecting the magnitude of pollution in the Great Lakes Basin were soil type, land use intensity and materials useage. For example, the intensive agricultural cities such as row cropping (e.g. growing corn, oats, soybeans and vegetables) on soils with fine textures (i.e. high clay content) contributed the greatest amounts of phosphorus. Areas of high phosphorus loading from intensive agricultural activities include northwestern Ohio and southwestern Ontario.

Thus, on the first two questions, PLUARG reported to the IJC that the boundary waters of the Great Lakes System were being polluted from non-point sources. At the time, PLUARG was doing the study, the words diffuse source and land drainage were, for present day terminology, the same as non-point source. Further, the report clearly identified agriculture as a major area where soil erosion could have serious off-site impact. However, many farmers were concerned about PLUARG's findings.

Many feared that if their operations were identified as sources of pollution they would be subject to regulation. Those who cooperated, voluntarily, with their governments to install soil and water conservation systems on their farms and forested areas, considered themselves as among the first environmentalists. They had long accepted the need to conserve and protect their basic resource, soil. The long-term programs, in the U.S., of the U.S. Department of Agriculture (USDA) agencies, namely the Soil Conservation Service (SCS) and the Agricultural Conservation and Stabilization service (ASCS), had been aimed at a reduction in soil loss. The efforts of these agencies supplemented by other USDA agencies such as the Forest Service and the Cooperative Extension Service, had resulted in significant reductions of certain types of soil erosion and indirectly had contributed to non-point source pollution control. However, this was not a primary purpose of USDA activities in research, extension, financing and technical assistance. Farmers and foresters were engaged in producing food and fiber, the more basic aspects of life, including growing crops, raising livestock and poultry, harvesting timber and other routine agricultural activities. The potential non-point sources were areas that appeared to many as part of a long familiar landscape, one very common to most citizens. They were the croplands, barnyards, woodlots and pastures in rural areas. The water in the streams could carry pollutants from these sources that had long been visible, but accepted as normal. They consisted

of soil eroding from crop fields, pastures and woodlots
and manure from barnyards.

With the passage of the Clean Water Act amendments of
1972, the relation of traditional soil conservation
efforts to water quality was more clearly defined.
During the PLUARG study, state and area wide planning
agencies were provided funds (from 1974-1980) to
prepare comprehensive water quality plans. Most of
these plans in the U.S. included agricultural non-point
source control elements. Two very significant non-
agricultural non-point sources, construction sites and
vast impervious surfaces such as streets, parking lots,
and roots, are mainly urban sources. Importantly,
concentrations of pollutants in runoff from non-point
sources can be higher than concentrations in municipal
wastewater. However, non-point source pollutants
usually flow directly into water bodies without any
treatment. However, during the PLUARG work, and even
through the present time, pollutants generated by non-
point sources are often highly complex and difficult to
track. A certain amount of non-point source runoff is
of natural origin. It is often difficult to separate
the impacts of point and non-point sources. Baseline
water quality information is lacking because monitoring
programs historically have been oriented toward point
sources of pollution. Cause-and-effect relationships
between non-point sources and particular water quality
problems are hard to establish due to the diffuse
nature of non-point runoff, and the many land use
activities within a given watershed. The performance
of appropriate management controls is highly dependent
on site-specific factors and is therefore difficult to
predict and assess. Lakes are the primary recipients
of these pollutants.

The third question, for many of the reasons already
cited, was obviously the most difficult for PLUARG to
answer. The answers to the two prior questions
required that a remedy or strategy for addressing non-
point source pollution, as a problem to the Great Lakes
Systems, be recommended to the Commission. PLUARG was
challenged by an attitude that non-point source
pollution programs would compete with point source
funding. The variety of disciplines, represented in
the reference group, had to resolve their differences
as to the recommended strategy. The tendency to accept
the strategy that would offer the least resistance to
adoption was apparent. PLUARG did not seek the answers
that would indicate resolution through the lowest
common denominator.

There was also concern that, based on cost of remedial
measure, any strategy would not deal adequately with

those periodic events that are extreme in nature. These include excessive precipitation, abnormal freezing and thawing of soils during snow melt, or high wind storms. PLUARG did caution IJC that the strategy for resolving non-point source problems would be neither simple nor inexpensively accomplished for both agricultural and urban areas.

Non-point sources of water pollution are characterized by their wide variety and large numbers of sources, the seemingly insignificant nature of their individual contributions, the damaging effect of their cumulative impact, the intermittent nature of their inputs, the complex set of natural processes acting to modify them and the variety of social and economic interactions involved in a proposed strategy to improve water quality by managing land use activities.

In the 1970s, environmental problem emphasis had largely been directed at point sources, which were more concentrated and under individual responsibility. Program administrators held the belief that non-point sources were either not controllable, or only so at large public expense. Also, there was a lack of an adequate legislative and regulatory basis for control.

PLUARG, in the recommendations to IJC, did not favor across-the-board measures for non-point source pollution control, but rather a methodology whereby problem areas are defined as a priority basis on which the most practical means for a particular source are then applied. The formulation of management plans would include a number of considerations that have not been comprehensively addressed in point source control programs of the past. Four major components were identified as follows:

1. information, education and technical
   assistance,
2. planning,
3. fiscal arrangements, and
4. regulation.

In addition, PLUARG recommended and the Commission agreed that the ecosystem, of which man is a major component, is complex and interdependent. All the potential impacts of man's activities need to be assessed and considered to the extent possible, when public decisions are made. Not all these impacts are obvious. They may appear much later, or in other parts of the system. This approach is important to rational resource management strategy for non-point source control.

The activities causing pollution, efforts to correct
such problems, and even seemingly unrelated policies
affecting land use activities, can have far reaching
consequences that may not be individually or readily
apparent.   In addition, the successful implementation
of any strategy will rely heavily on the concern,
interest and action of many individual members of
society.   PLUARG initiated citizen input to the study
by establishing a public information and consultation
program midway through the investigations.   There were
nine U.S. and eight Ontario public consultation panels,
representing a wide range of interests, including
agriculture, industry, environmental groups, education,
labor and local elected or appointed government
officials.   Their open meetings, panel-identified
problems and proposed solutions, and review of draft
documents were of great value in determining the
feasibility and practicality of the final
recommendations of the IJC.

PLUARG presented, as a primary recommendation, the
preparation of comprehensive management plans by the
respective jurisdictions, as an essential part of an
effective non-point source control program.   The
Commission, in their 1980 report to the governments of
the United States and Canada (IJC 1980) endorsed this
concept.   A management framework would be required
within which all governments can work in concert to
achieve coordinated, equitable and comprehensive action
within and between jurisdictions.   Within that
framework, remedial plans should be developed for
priority areas.   This was a cornerstone issue.

Further recommendations outlining essential elements of
a management framework or plan were given to the IJC.
This would enable them to provide necessary guidance
for individual jurisdictions in the design of their own
specific plans.   If needed, measures were urgent or
ongoing during the planning process, the
recommendations was to proceed on a priority basis.
Remedial measures, along with probable costs, were
developed for phosphorus runoff control from both urban
and rural non-point sources.   These were keyed for up
to three levels of measures and costs.   Questions about
the biological availability of phosphorus loads, both
over time (from year to year and seasonally) and
between nearshore and open lake areas required
additional work.   PLUARG submitted two supplementary
reports to the IJC in March and May of 1979.   Sound
conservation practices at minimal cost, that provided
long-term benefits to the land user, had been developed
to control soil erosion, for decades.   However, few of
these measures were developed specifically to reduce
water quality impacts of soil loss.   It was not the

land use, per se, that affects water quality, but rather how the land was managed. PLUARG, therefore, recommended that agencies assisting farmers adopt a general program to help land users develop and implement water quality plans. This program should include:

    1.a single plan formulated for each farm, where needed,

    2.consideration of all potential non-point source problems related to site-specific agricultural practices, including erosion, fertilizer and pesticide use, and livestock operations, and

    3.a plan that was economically viable.

Three major agricultural areas that were discussed in detail by PLUARG needed attention by the agencies formulating a general plan. These were soil erosion, livestock and poultry waste and commercial fertilizers. Heavy reliance was placed on education and the voluntary approach to the initial phases to demonstrate the need for accelerated erosion control and building on the stewardship ethic. However, the adequacy of existing and proposed legislation should be assessed to ensure there is a suitable basis for the enforcement of non-point pollution remedial measures, in the event that voluntary programs are ineffective. PLUARG felt that non-point programs of the future must include both voluntary and regulatory components, with the most emphasis placed on the preventative aspects of the laws and regulations directed toward control of non-point source problems related to agriculture. PLUARG also recommended the preservation of valuable wetlands and the retention for agricultural use of those farmlands that have the least natural natural limitations for farming. Within the Great Lakes Basin, there are many areas with unique features which should be retained to help reduce non-point source pollution. This usually requires governmental assistance.

POST PLUARG DEVELOPMENTS

In March of 1980, the IJC, in their report to Governments (Pollution in the Great Lakes Basin from Land Use Activities) had accepted PLUARG's findings for the most part, on the technical matters regarding sources and quantities of pollution. The Commission even expanded on PLUARGs concept of overall management plans and strategy.

The control of non-point water pollution associated with distinct land use activities will require increased involvement by existing agencies in the management of the problems. PLUARG's public consultation panels strongly opposed additional layers of government. They were concerned about too much existing government, about poor or non-existent coordination, both within and between levels of government. There was need for a concerted effort to minimize the overlap of programs and jurisdictions and to align government goals and objectives. This would require great effort. The recommended management plans were to be site-specific and were to include:

1. a timetable indicating program priorities for the implementation of PLUARG's recommendations,

2. agencies responsible for the implementation of programs designed to satisfy the PLUARG recommendations,

3. formal arrangements that were to be made to ensure coordination and cooperation,

4. the programs through which the recommendations will be implemented by federal, state and provincial levels of government,

5. sources of funding,

6. estimated reduction in pollution to be achieved,

7. estimated cost of these reductions, and

8. provision for public review.

In June, 1982, the U.S. General Accounting Office (GAO), issued a report to the Secretary of State stating that the U.S. Government had not adequately supported or been sufficiently involved in the water quality activities of the U.S./Canadian International Joint Commission, (GAO/CED-82-97). In 1981 the Great Lakes Water Quality Board had established a Non-Point Source Control Task Force to review and evaluate the effectiveness of PLUARG, IJC and other activities, in reducing non-point pollution during the prior five years. In its report to the Water Quality Board (IJC 1982), the Task Force made eleven recommendations. The

document strongly endorsed the earlier work of PLUARG
and the IJC.   Although extensive application of most
individual remedial practices had not yet occurred, it
was possible to identify some of the most successful
practices by examining the experiences of various U.S.
and Canadian watershed studies as summarized in Table
2.   The principal factor influencing the adoption of
agricultural control measures and practices is the
attitude and acceptance of the farmer.   Those practices
that are perceived to bring the most benefit to
agricultural production and/or profit, are most readily
adopted.   The Task Force contrasted the progress made
on implementation of point and non-point source
programs and stated that non-point implementation
difficulties included:

1.lack of clearly defined institutional
responsibilities,

2.almost total lack of funding arrangements,

3.reliance upon voluntary adoption of new
measures and practices by the rural farm
population, traditionally characterized by
individuality and conservation response,

4.technologies which, in many cases, are not
well demonstrated, either for ease or cost of
adoption and effectiveness in solving the
identified pollution problems,

5.lack of awareness and difficulty of showing
the relationship between water quality
problems and non-point sources which are
diffuse and periodic in nature, and

6.lack of a clearly defined source which can
be treated and upon which an agreed to
effluent standard can be applied.

These challenges to successful strategy for non-point
source pollution reduction again validates PLUARG's
concern that, because the physically and
institutionally diffuse and complex nature of non-point
source problems are real, there is a need for a
systematic approach on a wide scale.   This need for a
broad perspective leads logically to an ecosystem
approach for addressing non-point sources.   This
approach was reflected in the PLUARG report
"Environmental Strategy for the Great Lakes Basin" in
1978.   There is increasing evidence that the non-point
sources of pollution warrants national attention.
However, Lee M. Thomas, the Administrator of the
U.S.Environmental Protection Agency, has stated that:

...."States must take the lead in managing
non-point sources because they have the
adaptability, perspective and intimate
knowledge to develop such site-specific
solutions. They can easily reach individual
landowners and operators and help them change
the way they manage the land."...

This point of view has been debated and although there
is increased participation by non-federal levels of
government the fact is, pollution control is often
nothing but a cross-media transfer. The planet we live
on is a closed system. There is no such place as
"away" where we can throw things. There will be need
for continued national participation as we attempt to
prevent pollution of surface waters from non-point
agricultural sources. If we succeed in preventing
water from running off the farmland, we run the risk of
contaminating groundwater by letting it leak into the
water table. As the States in the Basin place heavy
reliance on the role of conservative tillage,
additional research is needed.

IS THERE A STRATEGY FOR POLLUTION GENERATED BY
AGRICULTURE?

A recent United States Plan, in concert with the states
of Indiana, Michigan, Ohio, Pennsylvania and New York,
for reducing pollution to Lakes Erie and Ontario, has
two major themes. Utilizing the work of PLUARG, the
IJC and others, along with new knowledge gained from
practical experience, the plan is: to achieve full
compliance with the regulatory point source discharge
permit limits, and to assist farmers in better managing
their soil and nutrient resources. The focus will be
to keep soil and nutrients on the land where they are
productive. Voluntary measures have proven effective
in demonstrations for non-point source reductions in
agriculture. If the review in 1990 shows lack of
progress, regulatory authority will be considered. The
non-point source element of the plan focuses upon the
management of crop residues to prevent soil erosion and
nutrient loss, and management of fertilizers and animal
wastes to minimize downstream pollution.

Heavy reliance is placed on accelerating the use by
farmers, in priority areas, of conservation tillage,
especially no-till. Soil testing and education will be
used to reduce any overuse of phosphorus fertilizers.
Construction of adequate animal waste facilities will

be promoted in priority problem areas. The one
management practice identified as having particular
merit in the control of phosphorus from agricultural
sources is less intensive tillage, especially no-till.
Conservation tillage retains part or all of the prior
crop residues on or near the soil surface. This
protects the land from wind and water damage, and is
increasingly being accepted by farmers.

In review, development of any strategy must follow a
certain path (Figure 2). A step-by-step process, now
being discussed in environmental circles would build on

TABLE 2   AGRICULTURAL CONTROL MEASURES AND PRACTICES

I.   MANAGERIAL PRACTICES
   A. Material
      1. Commercial Fertilizer and Livestock Manure
         Management
      2. Pesticide Management
      3. Remote Location of Livestock Facility from
         Water Course

   B. Conservation Tillage Practices
      4. Reduced Tillage Systems
      5. Ridge Plant Systems
      6. Zero Tillage Systems
      7. Timeliness of Tillage

II. VEGETATION
      8. Crop Rotation (sod-based)
      9. Contour and Strip Cropping
     10. Cover Crops
     11. Buffer Strips
     12. Windbreaks
     13. Double Cropping Systems

III. STRUCTURAL
     14. Grassed Waterway
     15. Terraces
     16. Surface Water Diversions
     17. Drop Inlet Structures
     18. Sediment Basin
     19. Stable Ditchbank Construction and Regular
         Maintenance
     20. Armoured Bank Protection
     21. Tile Drainage
     22. Livestock Manure Storage
     23. Livestock Manure Storage
     24. Excluded or Limited Livestock Access
         to Water Courses
     25. Adequate Control of Milkhouse Wastes
     26. Critical Area Planting

the experience gained during the "208" planning
activities of the 1970s. The strategy requires that
the problems be identified, that a set of goals be
established, and that a plan be designed that could
solve the problems and achieve the agreed to goals.

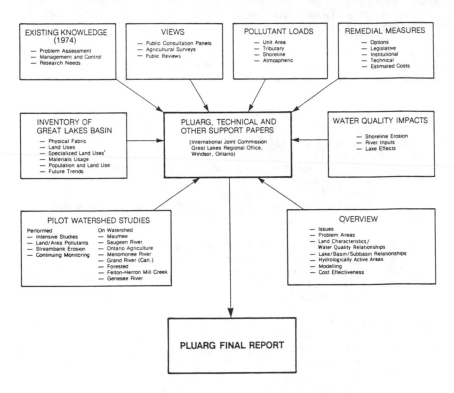

Figure 2.   Overview of PLUARG study activities.

A unique feature would be to test the strategy, by
implementing key components of the plan over a two or
three year time frame, with heavy reliance on
education, demonstration, and use of incentives (the
carrot), to convince land users to adopt non-point
source controls voluntarily.   If after a reasonable
time, the goals are not being met, the strategy would
be revised.   At this step in the process, the use of
regulation (the stick) would be strengthened.   The
objective would be this: as farmers want to plan for
and solve the pollution problems their way, to be self-
contained, they would have to face priority issues.

The potential problems areas must be given attention in the use of available, but limited resources. Target regions, as those labelled "hydrologically active areas" in PLUARG, would be inventoried and provided with the maximum of educational, technical and financial assistance. Plans for managing nutrients and the pesticides utilized in those areas would need to be developed and applied in an agreed to schedule. The state-of-the-art ("Best Management") practices would be used until research had improved technology. Tracking and monitoring of implementation would be accelerated. Research to add knowledge to non-point source pollution cause and effect relationships is needed.

SELECTED REFERENCES

Reports

IJC July 1978.   International Reference Group on Great Lakes Pollution from Land Use Activities. Environmental management strategy for the Great Lakes System.   1973 pp., IJC Regional Office, Windsor, Ontario, Canada.

IJC June 1978.   Great Lakes Research Advisory Board. The Ecosystems Approach. 47 pp., IJC Regional Office, Windsor, Ontario, Canada

IJC March 1980.   IJC Report to the Governments of the U.S. and Canada.   Pollution in the Great Lakes Basin from land use activities. 141 pp., IJC Regional Office, Windsor, Ontario, Canada.

IJC 1982.   Report to the Great Lakes Water Quality Board.   Non-point source abatement in the Great Lakes Basin - An overview of post-PLUARG developments. 129 pp. IJC Regional Office, Windsor, Ontario, Canada.

GAO 1982.   (U.S. Government Accounting Office). Report to the Secretary of State, International Joint Commission water quality activities need greater U.S. support and involvement.   16 pp. GAO/CED-82-97, June 23, 1982, Washington, D.C. 20548.

US EPA 1985.   1984 Report: National water quality inventory of the U.S. Congress.   77 pp. USEPA 440/4-85-029, Washington D.C. 20460.

Association of State and Interstate Water Pollution Control Administrators. Reauthorization of the Clean Water Act. 5 pp. Hall of States, 44 North Capitol Street N.W. Suite 330, Washington, D.C. 20001.

US EPA 1985. Task Force plan for phosphorus load reduction from non-point and point sources on Lake Erie, Lake Ontario and Saginaw Bay, October, 1985. 8 chapters. EPA, Chicago, Illinois.

NACD Executive Summary (draft), April 1986. A phosphorus diet for the Lower Lakes. NACD, Ft. Wayne, Indiana.

DNR January-February 1986. Non-point source pollution. Where to go with the flow. A Special Report. Madison, Wisconsin.

Natural Resources Council of the U.S. and The Royal Society of Canada. The Great Lakes Water Quality Agreement. An evolving instrument for ecosystem management. National Academy Press, 2101 Constitution Ave. N.W. Washington, D.C. 20418 or Royal Society of Canada, 344 Wellington, Ottawa, Ontario K1A 0N4

Journals

SCSA 1985. Non-point water pollution. A special issue. Journal of Soil and Water Conservation. January - February, 1985, 176 pp.

CHAPTER 15

INTRODUCING SCIENTIFIC ANALYSIS TO PUBLIC POLICY -
AN INTERNATIONAL PERSPECTIVE

Konrad von Moltke

Adjunct Professor, Dartmouth College, Nathan Smith
Hall, Hanover, NH 03755 USA
(Senior Fellow, The Conservation Foundation)

The relationship between scientific analysis and public
policy has evolved significantly since Max Weber's
famous essays "Wissenschaft als Beruf" (Research as a
Profession) and "Politik als Beruf" (Politics as a
Profession). They described an idyl which was outdated
when written, but have come to represent a canonical
text for the scientific community: the description of
a state of beatitude, of pure scientific being. Long
before Max Weber wrote his essay, science had become
inextricably intertwined if not with politics then at
the very least with policy - indeed, the essay was a
reaction against this distinctly discernible trend of
the 1920's and the experience of National Socialism
(and the attitude of the universities under National
Socialism) in the country of Max Weber has done little
to dim the glow of these essays.

Sadder and wiser through the experience of nuclear arms
and nuclear energy, of the research-military-industrial
complex, the very real dilemmas arising from the
relationship of science and policy have yet to be
overcome - neither can research remain in its pristine,
Weberian state nor can it degenerate into the hand
maiden of politics as which it appears in defense
appropriations for research.

In hardly any policy area are these dilemmas more acute
than in the field of environmental policy - with the

exception that they are not fraught with the agonizing moral component implicit in the relationship with other policy areas. Is not the scientist who rushes to warn of potential environmental hazards before they can be considered scientifically proven - one thinks immediately of the case of ozone layer depletion or forest dieback as examples - doing a public service? Or is he or she acting no differently from the scientist who sells new research findings to a defense contractor before reporting them in the research literature?

The relationship between research and environmental policy is intimate indeed: without research there would be no environmental policy as we now know it. Policy in relation to natural resources is unique among policy areas in its focus on goods which have no voice of their own. But policy is made by people, and through the articulation of goals and social means, hence to develop environmental policy it is essential to formulate the changes which occur in nature in an articulate manner. Short of impressionistic assessments - "this water tastes bad", "the linen comes off the washing line dirtied by smoke from that factory", "the trees are dying in my backyard", "tomatoes are not what they used to be" - the only means available for transforming environmental changes into articulate statements is - research.

It is consequently not surprising that researchers have played a key role in setting the environmental agenda, nor that environmental policy issues are spawning a substantial research effort in public and private scientific institutions around the world. Nevertheless the basic dilemma remains - how are research findings marshalled in a manner which allows policy conclusions to be drawn from them?

The solution to this dilemma is of significance to the course of environmental policy. It is a task of research and of politics alike - as any person who has been involved in such an exercise knows. The politics begin with the selection of institutions and of persons to undertake a policy-oriented assessment of research. As any politician knows, the essence of politics is the right choice of actors. Some may respond that in research objective criteria guide the choice of persons. This is the idyllic view of the process. It is true that the choice of persons must be justified in terms of objective criteria, but the sociology of science tells us much about the internal politics of the scientific community as it struggles to steer a course between objective criteria and personal or group prejudices, much as any parliament struggles to balance the public good with the human failings of the men and women who make up the institution.

There are many misunderstandings about the delicate balance between objective criteria and sociological factors contributing to policy-oriented research assessments. Many people appear to assume that research assessments of a given problem - the risk of ozone depletion, the causes of forest dieback, the hazards posed by a given chemical substance, the threat of extinction to a given species, health risks of air pollution, etc., etc. - should be universally valid, the true hallmark of science. In practice, taken as a whole, such assessments tend to be specifically oriented towards a specific time and place - the hallmark of politics - yet all the constituent parts represent objectivized elements.

It is therefore hardly surprising that every country goes about the business of interpreting research policy in a manner which reflects its particular political and administrative culture. At the extreme end it is possible to argue that this is even true of different jurisdictions within countries: it is certainly conceivable that a policy-oriented research assessment of the same issue in California could produce results different from an assessment of the same issue in Louisiana or Vermont.

Clearly the interpretation of research for policy is ultimately the task of policy-making institutions subject to pressures from many groups which will seek to provide policy-oriented research assessments of their own; the press frequently plays an important role in forming public opinion on the issues thus addressed. This paper will not discuss the details of this process - which merits an entire book - but rather address the formal institutions by which several countries marshall research findings for policy purposes and then proceed to discuss the implications for large lakes which typically involve a significant number of jurisdictions with potentially differing assessments of the same research evidence.

Traditionally, academies were the institutions where scientists met, exchanged ideas and developed a consensus on the major research issues of their time. The decades in some of the German academies of the 18th century can still make interesting reading, and the academies and Royal Societies were veritable battlegrounds of ideas in England in the 19th century. It is striking that these institutions have not generally adapted to the needs of policy-oriented assessments - their primary objective remains to serve the scientific community. This makes them institutionally uneasy about the ambiguities involved in such assessments.

In the United Kingdom, the Royal Society remains important in establishing the hierarchy among scientists but has not yet developed a particularly active role in environmental affairs, despite attempts on the part of some members to move it in that direction. A study on acidification and long range air pollution effects - financed by the Central Electricity Generating Board and undertaken together with more policy-oriented Scandinavian counterparts - is the major exception to this rule. The Royal Society has, however, participated in discussions about policies to limit the risks of hazardous chemicals and genetic research, but rather as an interest group committed to protecting the interests of researchers than as a mediator between research and policy.

In the United Kingdom, Royal Commissions can play a vital role in policy-oriented assessments. Frequently these are given a limited charge to consider a specific issue but in the environmental field, a permanent body, the Royal Commission for Environmental Pollution (RCEP), was established in the early 1970's and has acquired a respected place in the policy debate. Royal Commissions are not composed exclusively, or even predominantly, of scientists. In principle, there is no reason why there should be any scientists on a Royal Commission. In the environmental field, scientists have played a prominent role on the RCEP, and the chairmanship has thus far always fallen upon a distinguished scientist - distinguished among other things, by being a fellow of the Royal Society. It is therefore a body composed of both scientists and representatives of major interest groups and can take into account both published research and assessments, made by other institutions or interest groups, which are brought to it's attention. The reports of the Royal Commission do not aspire to be research assessments - although they do aspire to high scientific reliability - and can, and do, advocate specific policy options such as the banning of lead additives in petrol or the development of the "best practicable environment option" (BPEO) as a guiding principle of policy.

In comparison to the British Royal Society, its' Scandinavian counterparts can be described as activist. In political systems which pride themselves on their consultative processes and their ability to integrate all interests, the Royal Societies of Sweden and Norway represent the "scientific" point of view and have been pushed towards an active role in the formulation of environmental policy, for example on "acid rain" issues. The Swedish Royal Society publishes a journal, "Ambio" which reports on both research findings and current environmental policy issues. The Royal Society also undertakes policy research through it's own research institutes, such as the Beijer Institute. A peculiarity in the Swedish system are the Nobel symposia, linked to the Nobel Prize Foundation, which bring together researchers on important policy issues in a given political context.

The academies which were such a characteristic of German intellectual life in the 18th and 19th century continue to exist but have ceased to play such a prominent role. They were partially a response to the political fragmentation of Germany and did not make all the transitions of the 19th and 20th century, punctuated as they were by wars: first to centralization, then democracy, then dictatorship, the occupation and finally the establishment of two German states. The Berlin Academy of Sciences, from 1870 to 1945 a significant national institution, happened to be physically located in the eastern part of Berlin (where it continues to exist, and indeed to play an important role as do the academies in all Eastern European countries). For a variety of reasons, a parallel body was not set up in the Federal Republic of Germany - most importantly because of as initial unwillingness to undertake steps which might be construed as cementing the division of Germany into two states. Consequently the institutions of the scientific community in the FRG are a continuation of their predecessors in the Reich, Republic and Third Reich where they happened to be physically located in the western part of the country (as is the case for the former Kaiser Wilhelm Gesellschaft, now Max Planck Gesellschaft - MPG). In the other cases, institutions had to be newly "invented", as in the case of the Deutsche Forschungsgemeinschaft (DFG) which now acts as FRG partner of the British and Scandinavian Royal Societies. The DFG is essentially a research funding agency (which derives its support almost entirely from public funds) and has abstained from taking policy positions, except to defend the interests of researchers as a political pressure group. The DFG does, however, have a special panel for environmental research and has established a committee which defined maximum workplace concentrations of hazardous chemicals (MAK-Kommission). These maximum concentration values are of course important to laboratory workers, of whom the researchers funded by the DFG employ large numbers, as well as industrial workers and are exclusively research-based with no direct political effect.

Because of this relative lack of policy input, the MAK-
values of the DFG are one of the rare cases where
research assessments have achieved widespread
international acceptance.

The FRG has a unique institution in the area of
research-based policy assessments in the
Sachverstaendigenrat fuer Umweltfragen (Council of
Environmental Experts - SRU). Modelled on the West
German Council of Economic Advisors (which was in turn
modelled on the Unites States' Council of Economic
Advisors), the SRU is composed exclusively of
researchers, selected to provide a spread of expertise.
They are nominated by the Minister of the Interior who
is responsible for environmental protection at the
federal level. The SRU has its own budget and staff and
is physically located in the Federal Statistical
Office, an agency which reports to a different segment
of the Ministry of the Interior, thus ensuring both
administrative control and a certain measure of
independence. The SRU prepares reports on issues of
environmental policy - both on its own initiative and
at the specific request of the Minister of the
Interior. These reports are strictly research based
and tend to be technical in nature, but may include
specific policy recommendations. The issues addressed
are not research but policy issues, that is typically
questions such as "Opinion on the North Sea" rather
than "An Assessment of Pollution Loadings of the North
Sea".

The Netherlands has a political system which is based
on consultation and consensus. To some extent it is
comparable to Sweden. It differs from the Scandinavian
countries, however, in its much greater social and
political heterogeneity which leads to a much higher
level of political conflict which needs to be resolved
through consultative bodies in advance of political
decision-making. It is consequently not surprising
that the Netherlands has a plethora of consultative
bodies, more than any other country - at least in
relation to the size of the country. At the same time
it does not have a body devoted exclusively to the
preparation of policy-oriented research analyses for
the purposes of environmental issues.

Consultative councils were established in connection
with many of the sectoral environmental laws adopted in
the 1970's and these have slowly been consolidated into
a smaller number. In some instances, these councils
are predominantly research-oriented. Most important
among the Dutch consultative bodies in the
environmental field is the Centrale Raad voor
Milieuhygiene (Central Council for Environmental

Protection - CRMH) which is essentially designed to advise the government on legislation and policy and is not a research based body with an interest in environmental affairs. The body is, however, an addressee of research and regularly reviews the available research materials.

Furthermore, The Netherlands has a unique institution known as the Wetenschappelijke Raad voor het Regeringsbeleid (Scientific Council for Government Policy - WRR), a collegial body of distinguished scientists appointed on a strictly limited, not staggered term during which they are provided with a budget and can employ research staff to provide a scientific assessment of issues they consider important for policy. Because of the regular change in membership, over the years a wide range of subjects, including environmental ones, get taken up by this body. For example, a recent council focused on relations between the Netherlands and the Federal Republic of Germany and produced a report on Rhine pollution as part of this exercise.

As mentioned, the tradition of the academies continues to flourish in the countries of Eastern Europe, in many of which the Academy of Science also has institutes devoted only to research. Frequently these institutes have been the source of important environmental research assessments and come as close as any institution in a socialist country to providing independent analyses of major policy issues such as environmental degradation.

In addition to the countries mentioned, almost every country in Europe has developed procedures by which to provide a link between research and environmental policy. The Haut comite de l'environnement (Superior Environment Commission) and the Conseil national de la recherche scientifique (National Research Council - CNRS) in France, the special networking between universities, institutes and policy-makers in Belgium, the Consiglio Nazionale della Ricerca (National Research Council - CNR) and its institutes such as the Istituto per la Ricerca sul' Acqua (Institute for Water Research - IRSA) in Italy and the Athens Academy, are further examples.

A further case of special interest to the Great Lakes is the Royal Society of Canada which until recently was as hesitant about undertaking policy-related work as its British counterpart but has apparently changed its attitude in working more closely with comparable American institutions on environmental issues of joint interest such as the Great Lakes and "acid rain". In addition to the Royal Society of Canada has developed the environmental impact assessment procedure in such a manner that it constitutes research and assessment of a specific major project proposal.

The United States has been left last in this brief discussion because research assessments undertaken in the United States are clearly geared to the American policy process - but they often have an impact which can far transcend the national context, for the simple reason that they are undertaken in the United States.

In few other countries is the link between research and public policy as readily accepted as in the United States. The regular "business" of citizenship - consulting with the government, testifying before electing bodies and commissions, communication with the press and with citizen groups - are readily accepted as part of the tasks of the scientist. Consequently the avenues by which research results are communicated and interpreted for the policy-making process are legion in the United States. It is not at present possible to provide a satisfactory systematic characterization of this complex process so that it is necessary to focus on the institutions devoted specifically to policy-oriented research assessments.

While several institutions in the academy tradition exist in the United States and most of them are willing to take on policy relevancy research assessments, one institution clearly stands out among them. The National Academy of Science (NAS) was created in the 19th century by Congressional charter. It was - in marked contrast to its European counterparts - immediately endowed with the utilitarian traditions of the American policy. A condition of the charter was that the NAS was not only to provide a forum for eminent scientists to meet and exchange ideas but from the outset it was to provide scientific advice to policy makers. In this sense the NAS has affinities to the ideals of the landgrant colleges. The NAS has built a minor empire upon this aspect of its charter, creating an institution devoted specifically to advising government on a wide variety of scientific issues relevant to policy, almost always on a "pay as you go" basis. That is, the agency requesting the assessment must also finance the necessary work.

The NAS has fellows which like other academies, play a role in the advisory aspect of the Academy. On a given issue, panels which are not solely composed of NAS fellows are appointed. In fact, panels can be constituted with no fellows as members. While scientific credentials is an important criterion in

selecting panel members, members are often selected
with a view to representing different research
"affiliations", i.e. scientists working in
universities, in independent research centers,
industrial research or environmental research
organizations. The selection process thus ensures a
mix of policy views in addition to research excellence,
the rules of discourse however clearly require
scientific justification of statements.

The NAS has undertaken many research assessments of
environmental relevance, for example on chemicals in
the environment, air and water pollution, forest
dieback and lake acidification.

Assessments are always published. In no country are
these bodies allowed to make "private" communications
to policy-makers. This is imposed both by the
requirements of both the scientific community and the
policy process. In both arenas, open debate is
considered an essential guarantee of quality and due
process. Despite the universal practice of
publication, policy-oriented research assessments
circulate less internationally than might be expected.
Two principal reasons contribute to this state of
affairs. One, the scientific community does not
consider policy-oriented research assessments
scientific documents in the strict sense of the term.
As has already been mentioned, such assessments have
characteristics which make them specific to a
particular country and a particular stage of policy
development. Hence they may not be directly applicable
elsewhere. And two, language problems can contribute
to a lack of circulation. But even translation -
undertaken by several of the institutions in countries
where English is not the native language - has not
generally overcome this difficulty. The major
exception to this observation are assessments by the
National Academy of Sciences in the United States.
These can indeed receive wide circulation beyond their
country of origin, because they are in English, because
the American research community plays a leading role in
defining the research agenda of their disciplines,
because people all over the world are reasonably
familiar with American political processes, and because
the United States is the United States- the leading
power of our age.

In general it must be said, however, that it is
extremely difficult to obtain assessments of research
with respect to a transjurisdictional ecosystem - like
large lakes - because of the difference which may be
obtained in the various jurisdictions. The major
difficulty lies not in the absence of relevant

research, nor in language problems but in the fact that the policy-making institutions for large lakes are either indeterminate or poorly known. Consequently, it can be very difficult to make assessments which are truly policy relevant. Without a clearly defined well functioning policy-making context, it is also impossible to develop effective institutions for policy-relevant research assessments. All attempts at such institutional development at the international level have thus far had only modest success. The most important of the international institutions is the Scientific Committee on Problems of the Environment (SCOPE) which is in turn affiliated with the International Council of Scientific Unions (ICSU), a body in which the academies and research funding institutions of most countries are represented. The ICSU is in turn related to UNESCO. Thus far United States participation in ICSU has not been affected by its withdrawal from UNESCO because ICSU is not formally a UNESCO program. These institutions have addressed major environmental issues from a research perspective, but their reports are necessarily very slow in preparation and not very sharply focused on a policy-making context. They have not played a central role in the development on international environmental policy.

It is striking that the most developed international policy-making forum in environmental affairs - the European Community - has thus far not developed an institutional structure for policy-oriented research assessments. Several meetings have taken place on this score, but they have all proven inconclusive. Discussions with the European Science Foundation (ESF), a body which has the academies and research funding institutions of the Council of Europe countries (that is a much larger number than the EC) as its members have met with extreme reticence, based not only on the difference in membership of the EC and the ESF.

In practice, policy-oriented research assessments for international policy issues must be undertaken ad hoc, much as international policy-making itself occurs. Thus the two examples of cooperative ventures by academies - the joint venture of the British Royal Society and the Swedish academy and the assessment of the Great Lakes Water Quality Agreement by the NAS and the Canadian Royal Society - are probably examples of the kind of assessments which it will be necessary to undertake in advancing complex policy issues in transjurisdictional situations.

CHAPTER 16

SETTING PRIORITIES FOR GREAT LAKES ENVIRONMENTAL
QUALITY RESEARCH

Andrew Robertson

Ocean Assessments Division, National Oceanic and Atmos.
Admin., 11400 Rockville Pike, Rockville, MD 20852

ABSTRACT

This paper provides a brief overview of the U.S.
environmental quality research programs on the Great
Lakes that support the Great Lakes Water Quality
Agreement, discusses the mechanisms used in setting
priorities for these programs, and considers the lack
of influence of the International Joint Commission
(IJC) on the priorities setting process. Funding for
such programs is provided largely from four federal
agencies - the Environmental Protection Agency, the
National Oceanic and Atmospheric Administration, the
Fish and Wildlife Service, and the U.S. Army Corps of
Engineers. A four-tiered conceptual model is used to
portray the federal priority setting process. Although
the Great Lakes Water Quality Agreement assigns
responsibility to the IJC's Science Advisory Board
(SAB) to provide advice on the research needed in
support of the Agreement, there is no good mechanism at
present for such recommendations to influence this
process. It is suggested that the influence of the SAB
on the priorities setting process could be improved if
the directors of Great Lakes research programs and
others who are involved in the priorities setting
process were more actively involved in the developing
of the SAB's recommendations concerning research
priorities.

INTRODUCTION

By the 1960s it had become clear that human activities
in the Great Lakes Basin and especially the use of the
Great   Lakes   for   waste   disposal   was   causing
deterioration of the water quality of this vital
resource.   Canada and the United States recognized the
threat posed by uncontrolled use of these lakes and
signed the Great Lakes Water Quality Agreement in 1972
with the purpose of restoring and enhancing Great Lakes
water quality.   This Agreement, which was revised in
1978,   calls   for   the   two   countries   to   adopt   common
objectives and to develop and implement cooperative
programs and other measures to preserve the aquatic
ecosystem and achieve improved water quality throughout
the   Great   Lakes   System.   It   also   assigns   special
responsibilities and functions to the International
Joint Commission (IJC).   This bi-national group, which
was established by the Boundary Waters Treaty of 1909,
has   as   part   of   its   mandate   the   responsibility   to
consider questions and matters of difference related to
the U.S./Canada boundary waters that are referred to it
by the two governments.   Among the responsibilities
assigned to the IJC by the Agreement is to establish a
board, the Great Lakes Science Advisory Board (SAB), to
provide advice on research and other scientific matters
referred to it by the IJC.

The IJC, in establishing the SAB, charged it with
assessing the adequacy of Great Lakes research efforts
related   to   the   Agreement   and   with   identifying
priorities for, and helping coordinate this research.
Recently a bi-national review of the Agreement was
conducted by a committee of experts under the joint
auspices of the National Research Council of the United
States (NRC) and the Royal Society of Canada (RSC).
The   results   of   this   review   have   been   published
(NRC/RSC, 1985) and include an expression of doubt
concerning   the   effectiveness   of   the   SAB,   and   by
inference the IJC, influencing Great Lakes research
priorities and programs.

This   paper   provides   a   brief   overview   of   U.S.
environmental quality research programs on the Great
Lakes and then discusses the mechanisms used in setting
priorities   for   such   programs.   It   proposes   an
explanation as to why the recommendations from the SAB
and IJC do not seem to be effective influencing
decisions concerning these programs and suggests a way
to   increase   the   influence   of   these   recommendations.
The discussion is largely limited to consideration of
U.S. research programs as the author is much better
informed   on   how   decisions   are   made   concerning

priorities for these programs than for comparable Canadian ones. However, it is believed that the situation described for the United States is applicable in a general sense also to the Canadian Great Lakes research programs.

GREAT LAKES ENVIRONMENTAL RESEARCH PROGRAMS

In the United States both federal and non-federal agencies support Great Lakes environmental quality research and related monitoring programs that are of potential value in support of the Agreement. Using a summary of funding levels for federal marine pollution research and monitoring programs during Fiscal Year (FY) 1983 compiled by the National Oceanic and Atmospheric Administration's (NOAA) National Marine Pollution Program Office (1984a, 1984b), it can be calculated that total spending for such programs was approximately $10.3 million during that year. The same office (National Marine Pollution Program Office, 1985a) has made a comparable estimate of $6.2 million for funding of non-federal marine pollution research and monitoring programs in the Great Lakes during the same fiscal year. However, when these non-federal funding levels are analyzed further (Table 1), it is found that approximately two-thirds ($4.2 million) of the funds were spent by private utility companies for monitoring the specific environmental effects of construction and operation of their power plants located in the coastal zone. Such monitoring has very limited applicability in support of the Agreement. Of the remaining $2.0 million almost one-half ($0.9 million) was spent by state and academic organizations as matching funds in conjunction with programs funded primarily with federal dollars. Thus, independent, non-federal funding for Great Lakes environmental quality research programs with clear applicability to the Agreement is relatively minor, having been no more than $1.0 million in FY 1983. These latter expenditures went almost entirely for support of environmental monitoring programs conducted by the states in their coastal waters.

The remainder of the discussion in this paper will focus on federal programs and the establishment of federal priorities because it is at this level where most of the funds are provided and most of the priorities are set for Great Lakes research.

TABLE 1 NON-FEDERAL FUNDING FOR GREAT LAKES RESEARCH
         AND MONITORING FOR FY 1983

	Millions of Dollars	
STATE AND LOCAL GOVERNMENTS		
Matching funds	0.7	
Other	1.1	
Sub-total		1.8
ACADEMIC ORGANIZATIONS		
Matching funds (Sea Grant)	0.2	
Other	0.1	
Sub-total		0.3
PRIVATE ORGANIZATIONS		
Utility Companies	4.0	
Other	0.2	
Sub-total		_4.2_
TOTAL		6.3

Four federal agencies provide support for research
programs with potential applicability in support of the
Agreement.   These agencies and estimates of their
funding for Great Lakes research in *fy 1985 are shown
in Table 2. These estimates are based on figures
compiled by the National Marine Pollution program
Office (1985b, 1985c) supplemented by information
obtained by consultation with the managers of these
programs. Two levels of funding are presented in this
table. The first column shows funds provided for Great
Lakes pollution studies. These are, in large measure,
directly applicable to support the Agreement. The
second column is included mainly to help put the
pollution expenditures in perspective and shows the
funds spent for all Great Lakes environmental research
and monitoring including the pollution studies. This
latter category includes programs, especially ones
related to the development of fisheries and other
resources, that have little direct applicability to the
Agreement.

NOAA's Great Lakes Environmental Research Laboratory
(GLERL) in Ann Arbor, Michigan, was the organization
with the largest amount of funding ($2.6 million) for
pollution research and monitoring in FY 1985.   This
laboratory conducts a research program on the physics,
chemistry and biology of the Great Lakes with the goal
of providing information to resource managers to help
maintain and improve the quality and usefulness of the
Great Lakes. The major part of the research conducted

TABLE 2   FEDERAL FUNDING FOR GREAT LAKES RESEARCH AND
           MONITORING FOR MONITORING FOR FY 1985

AGENCY	POLLUTION PROGRAMS	ENVIRONMENTAL PROGRAMS
	(Millions of Dollars)	
NATIONAL OCEANIC & ATMOSPHERIC ADMINISTRATION		
Sea Grant	0.8	4.9
Great Lakes Env. Res. Lab.	2.6	3.8
ENVIRONMENTAL PROTECTION AGENCY		
Large Lakes Res. Station	2.5	2.5
Great Lakes Nat.Prog.Off.	2.5	4.1
U.S. FISH & WILDLIFE SERVICE		
Great Lakes Fishery Lab.	0.7	3.0
U.S. ARMY CORPS OF ENGINEERS		
North Central Division	2.1	2.1

at GLERL consists of process studies to improve our
understanding and predictive capabilities concerning
the impact of pollutants on the Lakes. This pollution
research especially focuses on studies concerning
nutrients and toxic synthetic organic compounds. The
Sea Grant of NOAA supported a small pollution program
($0.8 million) with similar objectives at several
universities in the Great Lakes region in FY 1985.

Two subdivisions of the Environmental Protection Agency
(EPA), the Large Lakes Research Station (LLRS) in
Grosse Ile, Michigan, and the Great Lakes National
Program Office (GLNPO) in Chicago, Illinois, each had
FY 1985 pollution programs almost as large ($2.5
million) as GLERL's. LLRS conducts a research program
to provide technical support for the Water Quality
Agreement in such areas as water quality criteria,
eutrophication, and risk assessments for toxic
pollutants. It is part of EPA's Office of Research and
Development. GLNPO, in contrast, has a role more
closely allied to the regional regulatory activities of
EPA. It was established in EPA's Region V headquarters
in Chicago to focus attention on the significant and
complex natural resource represented by the Great
Lakes. Its primary goal is to develop programs,
practices and technology in order to eliminate or
reduce, to the maximum extent practicable, the
discharge of pollutants into the Great Lakes System.
It has responsibility for leading and coordinating U.S.
actions, including research and monitoring, in support
of the Agreement.

The Department of Interior's Fish and Wildlife Service also operates a laboratory which conducts research concerning the Great Lakes. This laboratory, the Great Lakes Fishery Laboratory in Ann Arbor, Michigan, is primarily involved with fisheries studies on aspects, such as stock assessment, that are of only limited applicability for support of the Agreement. However there is a relatively small pollution program ($0.7 million) at this laboratory which focuses on contaminant studies related to important sport and commercial fish species and their forage base.

The final federal research and monitoring program on the Great Lakes that has applicability for the Agreement is that conducted by the North Central Division of the U.S. Army Corps of Engineers in Chicago, Illinois. This organization does not normally support separate environmental research projects, but does support monitoring and data acquisition and interpretation as well as project-specific developmental research related to its operation activities. The magnitude of such work in the North Central District has generally been quite limited (less than $0.7 million annually) during recent years. In FY 1985, however, the North Central District spent a substantially larger amount ($2.1 million) primarily as part of a special demonstration study of an innovative management strategy for disposal of contaminated sediments. This study was focused on the Indiana Harbor channels at the south end of Lake Michigan.

SETTING GREAT LAKES RESEARCH PRIORITIES

Of course, the procedures used to set research and monitoring priorities vary in specific details among these federal agencies. However the general structure of the priority setting process is quite similar for all of them. The following discussion presents a generalized description of this process.

One point that the priority setting processes in the different organizations have in common is that they tend to be quite complicated. They involve decisions made at several hierarchical levels of management and by many different legislators, managers, and scientists. Each level in the hierarchy makes decisions that place constraints or limits within which the level immediately below must function. The discussion presented here uses a four-level conceptual model of this hierarchical federal structure to simplify the presentation. Although this model obviously does not furnish a completely accurate picture of the complexity of the federal priority setting process, it is believed to be representative enough of the true situation to facilitate presentation of the overview called for in this paper.

Obviously the **top level** in the hierarchy for all the agencies involves decisions by the U.S. Congress which determine the general directions for environmental research and monitoring supported by the federal government. The decisions made at this national level determine the priorities for funding the national environmental programs conducted by the various federal agencies. This is also the level where the opinions of the general public have their major impact. In some cases, the decisions at this level specifically establish priorities for research and monitoring on Great Lakes environmental problems. More often, however, the decisions at this level only deal with broad national problems, although these may have a Great Lakes element among their components. The priorities for the components within these programs are determined at the next lower level of the hierarchy.

Priorities at this **second** or **agency level**, as to which of the programs within an agency are most important are determined by the agency's top-level managers. Within the broad guidance from Congress, these managers set priorities among the competing demands of existing and proposed programs directed at specific environmental and related problems. It is at this level that the Great Lakes research laboratories and other research-related federal organizations usually compete for support against comparable units within the same agency that focus on other problems. Decisions at this level have a great effect in determining the magnitude and general direction of the studies that are carried out on Great Lakes environmental problems.

The **third level** in the priorities setting process is where priorities are established for the research projects needed to implement an agency's environmental problems. Decisions at this level do not involve setting priorities concerning which environmental problems (e.g. nutrient over enrichment, PCB contamination) to attack as in the preceding level, but instead to involve determinations as to which aspects of these problems (e.g., measurement of atmospheric inputs, determination of acute toxicities) demand the most immediate attention. Decisions at this level involve a balance of managerial and scientific considerations. With regard to Great Lakes programs, they are often made by the directors and other science program managers that direct the federal laboratories and research offices involved with Great Lakes environmental research and monitoring.

The **final level** in the four-tiered structure is where priorities are established for specific scientific tasks.   Here is where the scientists directing a project decide exactly which activities to carry out to meet the objectives of the study effectively and within the constraints of the resources available.   The decisions made at this level are based primarily on scientific considerations and so are relatively unimportant for the purposes of this paper.

The conceptual model of the priorities setting process for federal environmental research and monitoring emphasizes top-down decision making and this is, in fact, largely how the process operates. However, it is also important to realize that each level makes recommendations for priorities to the level immediately above it and that this bottom-up process plays a major role in determining the priorities at these higher levels. The recommendations that go up the chain usually include specific information supporting the proposals for programmatic and budgetary actions.   As this and other information provided from below often provides much of the factual basis for reaching decisions, what a lower level includes in its recommendations usually carries great weight in influencing the higher level decisions.

THE ROLE OF THE SAB IN THE PRIORITY SETTING PROCESS

The SAB has been charged to make recommendations for priorities in research and monitoring to support the Water Quality Agreement.   This is a vital role in furthering the Agreement as the SAB is the only formal mechanism whereby the two nations cooperate to develop such recommendations for the Great Lakes System as an entity and without regard to individual agency and national considerations.

Yet, as pointed out earlier, there are substantial doubts concerning the influence of the recommendations on research priorities developed by the SAB.   As the actual research in support of the Agreement is supported primarily by agencies in the national governments of the two countries, this means there is doubt concerning the influence of SAB's recommendations, on the decisions concerning research priorities, in these two agencies.

Why is this so? A comparison between the process by which SAB recommendations are developed and forwarded to the two national governments and the priority setting process described in the preceding section seems to provide much of the answer. Although it commonly draws on scientific and other expertise from both within and outside of the Great Lakes research community, the Board develops its recommendations rather independently of the federal priority setting processes of the two governments. In recent years the SAB has been composed of persons with widely varying backgrounds and experience related to the Great Lakes and representing a wide spectrum of the interests concerned with Great Lakes problems. Experts with experience working with academic organizations, industries and environmental interests have been well represented. There have been few members, however, who represent the major Canadian and U.S. federal research and monitoring organizations on the Great Lakes and participate in the federal priority setting process.

The recommendations of the SAB concerning Great Lakes research and monitoring are presented to the International Joint Commission which considers them and revises and forwards those that are approved to the national governments. It seems to be at this point that the recommendations tend to stall. After the IJC has forwarded its recommendations to the governments, there does not appear to be a good mechanism to link them to the priority setting processes within these governments. This lack of linkage provides both the cause and the answer to the problem of lack of SAB and IJC influence on the setting of Great Lakes research priorities.

If the SAB deliberations could be integrated in a better fashion into the process of priority setting in the two countries then the recommendations of this board would have more influence. This could be done if the Great Lakes research directors and others who are involved in the national priority setting processes, especially in the third tier of these processes, were more actively involved in developing the SAB recommendations for research priorities. Thus, it is proposed that the SAB bring such experts more actively into their deliberations. The SAB could serve as a forum where the Great Lakes research directors of the two countries could meet and attempt to develop consensus on the most pressing needs for research and monitoring in support of the Agreement. If such consensus could be reached then these directors could be expected to push within their own agencies and governments for programs to meet the identified needs. It may even be possible to reach agreement on which agencies should take the lead for conducting activities directed at specific identified high priority needs. In this fashion, the SAB could provide a mechanism to develop bi-nationally coordinated research programs that would increase the efficiency and effectiveness of research being conducted by the two countries in support of the Agreement.

ACKNOWLEDGEMENTS

I would like to acknowledge the contribution of Mark Monaco in assisting with compilation of the funding estimates used in this paper.

REFERENCES

National Marine Pollution Program Office 1984a. National Marine Pollution Program. Agency Program Summaries. FY 1983 update. National Oceanic and Atmospheric Administration, U.S. Department of Commerce, Washington, D.C.

National Marine Pollution Program Office 1984b. National Marine Pollution Program. Catalogue of Federal Projects. FY 1983 update. National Oceanic and Atmospheric Administration, U.S. Department of Commerce, Washington, D.C.

National Marine Pollution Program Office 1985a. Inventory of non-federally funded marine pollution research, development and monitoring activities - Great Lakes Region. National Oceanic and Atmospheric Administration, U.S. Department of Commerce, Washington, D.C.

National Marine Pollution Program Office 1985b. National Marine Pollution Program. Agency Program Summaries. FY 1984 update. National Oceanic and Atmospheric Administration, U.S. Department of Commerce, Washington, D.C.

National   Marine   Pollution   Program   Office   1985c.
National Marine Pollution Program.   Catalogue of
Federal Projects.   FY 1984 update.   National Oceanic
and Atmospheric Administration, U.S. Department of
Commerce, Washington, D.C.

National Research Council of the United States and
Royal Society of Canada. 1985.   The Great Lakes Water
Quality Agreement.   An evolving instrument for
ecosystem   management.   National   Academy   Press.
Washington, DC.

CHAPTER 17

THE ROLE OF THE NEWS MEDIA IN DETERMINING PUBLIC POLICY FOR THE GREAT LAKES

Michael Keating

Environmental Reporter, The Globe and Mail, 44 Front Street West, Toronto, Canada

ABSTRACT

The news media have played a catalytic role in shaping public policy on Great Lakes pollution issues, from the days of the eutrophication debates of the 1960s to the toxics problems of recent years. By publishing information about the deteriorating state of the lakes, the media created pressures for action on the part of government, industry and non-governmental organizations, the three main actors in Great Lakes environmental debates. To some degree the media are simply the method by which the main actors communicate with the public. But the media can also affect the content and impact of the message by giving more attention to some pieces of information and less to others. Once issues have been raised the media can spur debate by focusing continuing attention on them and demanding that officials respond to their questions. In fact governments have often found themselves in the awkward position of having raised public concerns about such problems of drinking water but then are unable to explain what they are going to do about those problems. There are now signs that some governments are re-thinking their strategies and may release less information which raises questions that they cannot answer.

You have all seen the headlines:

PCB Poisons Build Up In The Body

City Could Become PCB Dump, Mayor Asserts

Toxic Wastes: Tip Of The Iceberg

**Town   Fights   For   Plan   For   Toxic   Waste
Treatment Plant**

Industries Cited For Dumping Hazardous Wastes

**Lack Of Planning For Disposal Crisis May Lead
To Shut-Downs In Industry**

These are just a few headlines culled from my files.
And they come not from what is called the sensational
press but from such journals as The Globe and Mail,
including the business section, the New York Times and
New Scientist magazine of Great Britain.   They tell us
that the public and people in the news media are
concerned about the state of our environment and do not
feel that those in charge are, in general, coping
adequately with the situation.

In this paper, I will deal with the role of the media
in the shaping of public policy in the context of
toxins in the Great Lakes.   To give you some background
let me talk a little bit about the news media and how
they operate.   I will leave you to draw your own
conclusions about whether or not we are doing a good or
even adequate job.

I am a newspaper reporter and I will talk about
newspapers but much of what I will say about news
gathering applies to the electronic media.

WHO DOES WHAT INSIDE A NEWSPAPER?

The news story is typically written by a reporter such
as myself.   That story is filed with a middle level
editor who may order it reduced in size to fit a
limited amount of space.   The story goes to a copy
editor who reads it for style, grammar, writes the
headline and may do the trimming to fit a given news
hole.

In many cases the story is based on information
released during the same day it appears in print and
the story must be written and edited within a few
hours.

News stories may be written by salaried staff
reporters, such as myself, freelance writers who cover
regions where we lack staff and from the news services.
These services transmit stories from their own
reporters or from other news media around the world in
a giant news pooling operation.

In the paper there is a chain of command which judges
which stories will be assigned and which will be used.
This chain includes a series of copy, assignment, page,
section, metro, national, foreign, assistant managing
and managing editors.

In addition to the news reporters such as myself we
have editorial writers producing opinion pieces which
give the editor's viewpoint about what should be done
in a given situation.

What About the News Content?

We have been called an imperfect mirror of the world.

We have been accused of sensationalism, fear-mongering,
scare headlines, misquotes, inaccurate and incomplete
stories.

How do we choose what we call news?

Vast amounts of material, tens of thousands of words a
day, flow into the offices of a large newspaper so we
are highly selective in what we choose to print.

It is not a precise science. For example a story which
would get good coverage one day will get bumped to the
back pages on another day because of other major,
breaking news events such as an air crash,
assassination or collapse of a government.

HOW DO THE NEWS MEDIA APPROACH ISSUES?

The media often cast themselves in the role of an
unofficial opposition critic to institutions and there
is something of a tradition of standing up for the
underdog in conflicts.

And yes, our pages are full of conflicts which is not
surprising given that our historical role has been to

challenge ideas with other ideas and to act as a forum
for different viewpoints.

The reasons that we are such a good forum is that we
are apart from the other players in, for example, the
environment field. Since we reporters are not part of
government we are not confined to working within the
hierarchy and a reporter can pick up the phone and call
a cabinet minister directly. We are not constrained by
diplomatic niceties so we can call the minister in
another province, state or country.

Since we are not part of the "system" in the formal
sense we can call anyone for comments, including
politicians, polluters and environment groups. People
may not always respond but more often than not they do
feel an obligation to offer some kind of explanation to
our questions.

It is not usually the role of journalism to settle an
argument. That is done by the parties whom we quote.
The media acts as a marketplace for opinions.

This raises the issues of outside opinions getting into
print in their own words. There are two outlets for
people who would rather say it themselves. The first
is the well-known letters to the editor section, which
should be fertile ground for opinions, dissenting or
otherwise and should be a source of new information. A
larger canvas is provided in many papers in the form of
a block of space, often on the page opposite the
editorial page, for informed outsiders to write longer
opinion pieces. At the Globe and Mail we have had
several such articles written by experts on Great Lakes
water quality.

Let me finish off the mini-lesson on journalism by
talking about my specialty - environmental reporting.
This is a field of reporting which is coming of age as
environmental issues climb higher and higher on the
public agenda. Like governments the media are learning
that environment issues are more than crises to be
responded to by general assignment reporters. Slowly a
cadre of experienced writers is being formed.

MEDIA'S ROLE IN DETERMINING PUBLIC POLICY

Now let us get to the main issue, that of the role we
in the news media play in determining public policy.
For starters let me grab just one set of statistics out
of the hat. During last spring's election in Ontario

an opinion poll found that environment, couched in the terms of air and water pollution was the number four issue on voters' minds. It came after economy, jobs and health but ahead of education.

How did it get there?

Well many people can make their own observations of smoky skies over major cities and unnatural tasting drinking water but I would say that the majority of people form their opinions directly or indirectly from news stories. I add the word indirectly because in many cases we all obtain information second-hand from people who did get it from the news media.

To be more specific, what role do the media play in setting the political agenda?

As long as there is a free press and governments to report on, people will argue over whether the press sets the agenda for governments or governments manipulate the press. I have had cabinet ministers tell me their agenda and that of much of cabinet is set by what is on the front page of the morning paper. And I have seen environmental departments strive to achieve certain goals by feeding certain information to the media and suppressing or at least downplaying other information.

For example in the first case we might publish a story quoting a government spokesman, an outside research scientist, professor or an environmental group on a sensitive issue such as drinking water safety. The issue may then be picked up by an opposition critic and put to the environment minister during public debate. The minister must be able to respond because the news media are watching to see whether or not he or she appears to be on top of the issue.

This process can set off a sort of domino effect. The minister will ask his staff to research the subject. Civil servants will have to cull their files and possibly go to outside sources for information. The minister may make a statement promising action which will involve many researchers and may result in the construction of sewage treatment projects. Or the minister may criticize the source of information or another player, possibly even another government, resulting in a heated public debate and possibly diplomatic repercussions.

In the second case, that of media manipulation, a politician or even a senior official, may decide to mount a campaign against an issue such as acid rain or lead in gasoline.

Let's take the case of the minister. This persons decides the goal is a cleanup of a substance but is not certain that there is adequate public awareness and therefore support for the cause. The politician needs such support, often manifested in the form of letters and calls to the government or polling results, in order to get funds or changes in laws. The politician orders the department to pull together research studies which indicate there is a problem.

The studies are presented to the media who then respond by publishing stories quoting the minister and the officials who are presented as spokespersons. If the issue gathers momentum, either because of shrewd marketing or because it touches a nerve in the body politic you have prepared the ground for a policy change. The politician has used the media to arouse the public who then call on government to implement change.

The converse is true. A government department may have discovered a serious source of pollution of the Great Lakes but for various reasons does not want this made public. The reasons can vary. The politicians may want to protect business interests which support the government with donations. The government may not want to create a public demand for expensive clean-up programs which they would have to fund at a time when they are trying to reduce government spending. Or, they simply may not want to appear to lack control of the environment.

What I have said about the relationship between media and governments and the setting of agendas is true for many other fields, such as science, academe and business. Science has certainly benefitted from a heated discussion of environmental issues because it is to the scientists to whom we must all turn for answers.

The business community has certainly been affected by stories identifying industry in general and many specific businesses in particular as sources of pollution. Certain chemicals have been banned or severely restricted and to a greater or lesser degree debate about them in the news media was instrumental in those decisions. So we in the media are both a driving and a driven force. We are driven to publish certain things because we cannot ignore the statements of certain people. We are driving because we raise issues and demand responses. We are part of an information cycle which begins with a source, moves to publication, which is where we come in, then generates public awareness and demands for action and finally leads to action by those in charge.

For those of you with a scientific bent, I would offer the image of the news media as a catalyst. We help to cause a reaction but we are not the source of the raw material, the information, which is the basis of the reaction.

News Media as a Catalyst

The news media have certainly played a major catalytic role in the evolution of public policy regarding the Great Lakes. Since at least the 1960s, photographs of dead fish, ugly sewer discharges, soupy and soapy lakes and rivers have shocked the public and politicians alike. The media confronted the world with the ugly images of the problems created by a careless industrial society. We recited the facts and figures of rivers on fire and beaches unfit for swimming. We wrote of dead zones in Lake Erie and fish which are unsafe to eat and of toxic chemicals in the drinking water.

First it was the eutrophication problem, the so-called dying of Lake Erie and a similar threat to Lake Ontario. Then came the toxic substances issue even before phosphorus controls were completely in hand. The toxic substances problem in the Great Lakes became a major public issue in 1970 with the discovery of mercury in the fish of Lake St. Clair and adjacent waters. In the ensuing years there was growing concern as such chemicals as PCBs, Mirex, Lindane and Dieldrin were added to the list of dangerous substances found in Great Lakes fish.

Concern about the chemical threat to drinking water, at least in the Lake Ontario context, began in 1980 with the discovery that the highest levels of TCDD dioxin in herring gulls from the Great Lakes were from Lake Ontario. The description of TCDD as a super-toxic which could kill even in tiny amounts really sparked public concern. Citizens phoned newspaper reporters and government officials asking if they should drink the water or if they should even have children. Unfortunately there were no simple answers about the risk then and there are still none today.

At the same time the SCA pipeline debate was gathering momentum. This is a pipe to carry treated chemical waste into the lower reaches of the already heavily

polluted Niagara River.   There were many news stories
quoting SCA officials and some government officials as
saying there was no risk and there were stories quoting
some scientists and citizens as saying they did not
believe these assurances.

The issue of the big Niagara River dumps, Love Canal,
Hyde Park, S-Area and 102nd Street, leaking into the
river added fresh fuel to the already fiery debate.
Report after report talked about how some of the most
toxic chemicals around lay buried at the edge of the
source of drinking water for 8 million people in two
countries.   Story after story quoted environmentalists,
civil servants and politicians talking about who was
responsible and who should be doing what to clean up
the mess.

The hundreds of stories have had an effect at a local,
provincial and state, national and international level.
They have resulted in concern not only in the general
public but among scientists, government officials and
politicians, who rely on the news media for tipoffs
about upcoming problems and to get a sense of the
public mood.

Decision Maker's Response to Media Raised Issues

To be somewhat simplistic I would say there are two
sorts of responses by decision makers to the issues
that have been raised in the media.

Activist ministers and civil servants tend to react
well to a free flow of information.   They recognize
that each new revelation presents them with a challenge
but they are confident in their own ability to respond
in an intelligent way.   Even if they do not have
instant solutions they have goals and can articulate
them.   As a result they encourage the release of
information from their own departments.

Reactive governments are those on the defensive and
they react poorly to the free flow of information
because they see it as something which raises questions
they cannot or will not answer.   They fear the fast-
moving debate which results from many people having
access to raw data fresh from their laboratories.
These governments try to reduce the flow of new
information, preferring to endlessly recycle what is
already in the public domain.

Where Do We Go From Here?

The public debate about the state and the fate of the
lakes is well under way. More and more news media are
taking a sustained and informed interest in the issues.
But we in the media rely on you in the broad field of
environment information to keep us on the right track.
I encourage you all to assist in that process.

CHAPTER 18

THE ROLE OF RESEARCH IN MANAGEMENT OF THE LAURENTIAN GREAT LAKES

Alfred M. Beeton

Great Lakes and Marine Centre, University of Michigan, Ann Arbor, MI

INTRODUCTION

We need results of sound research on large lakes because of their immense importance as resources of great aesthetic, environmental, and economic value. Effective management cannot be realized without the knowledge provided by such research. Furthermore, large lakes provide opportunities for research into fundamental properties and processes of aquatic ecosystems. The great lakes of the world, because of their size, have characteristics which make them intermediate between small lakes and marine oceanic bodies. These characteristics lend themselves to study of gradients in physical, chemical, and biotic interactions found in large bays and in the coastal and offshore regions.

The inshore/offshore differences favor studies of frontal processes such as thermal bars. These lakes are of sufficient size to affect the climate of a region and yet their drainage basins have a lesser effect on internal processes than in small lakes. The differences found among the St. Lawrence Great Lakes provide opportunities for comparative studies of many variables over a broad range of environments. Detailed studies can be made of ecosystem processes in physically bounded and clearly delimited basins. The sedimentary environments of large lakes are especially favorable for paleolimnological studies. Physical processes are of much greater importance in large lakes than in small, e.g. separate water masses occur, upwelling and sinking are important, and they are visibly affected by the Coriolis Force.

A large amount of research has been conducted on the Great Lakes, most of it during the past 30 years. We now have reasonably good comprehension of many aspects of the ecosystem and we are effectively dealing with truly horrendous crisis such as the sea lamprey and eutrophication. Only 30 years ago, a list of planktonic species in Lake Huron would have been an original contribution! Nevertheless, in spite of the progress which has been made, we are not in a position to provide the kinds of information necessary to manage these large lakes. The reason is that much of the research has been in response to crisis management as seen by the history of research.

HISTORY OF RESEARCH

Applied Research

The history of research on the Laurentian Great Lakes is closely tied to their uses. The earliest use of these lakes by European immigrants was transportation. The lakes and their connecting channels and tributaries made the heartland of North America readily accessible. It is not surprising that the earliest recorded observations of the lakes were of level fluctuations as commented on by Father Marquette in 1673, Father Louis Hennepin in 1679, Baron Horton in 1689, and Charlevoix in 1721. By 1860, the United States Government had established water-level gages at a number of harbors.

The abundant supply of fish played an important role in the settlement and development of the Great Lakes region. Fish provided an inexpensive and wholesome food supply. Before the early settlers, the Indians of the region lived largely on fish and tribes made periodic migration to lay in stores of fish for winter use (State Board 1867). The early accounts state the fish were so plentiful that the supply was considered inexhaustible. The commercial fisheries of the early 1830s were mainly in the Detroit, St. Clair and St. Marys rivers, Straits of Mackinac, southeastern Lake Superior, and Saginaw Bay. It is interesting to note that catch statistics for 1836 and 1837 indicate that the Detroit River fishery produced about one-half of

the total pounds of fish taken in the Great Lakes in
those years. Almost all the catch was whitefish
(Bissell 1887). The introduction of more efficient
fishing methods soon began to have its effect. For
example, a fisherman at Mackinac reported that he
caught more fish in 1843 with 12 gill nets than he
could take with 240 gill nets in the 1880s. Whitefish
were becoming scarce in the rivers by 1871. The
decline in the fishery was attributed to three factors
(State Board 1887):

1. fishing with small mesh nets which
destroyed young fish,

2. pollution from sawmill refuse and fish
offal, and

3. fishing for spawning fish.

The concern over declining fish stocks led to early
fisheries research (primarily surveys), restrictive
legislation and construction of a number of fish
hatcheries in Canada and the United States.

These early efforts to research and manage fish stocks
were generally ill-advised and fish populations
continued to decline. The collapse of the cisco
(Coregonus artedii) fishery, which had produced up to
40 million pounds annually in Lake Erie resulted in two
major limnological studies of this lake in the late
1920s and early 1930s. The reports of these studies
recognized deterioration in the nearshore waters as a
consequence of pollution, but they failed to recognize
the importance of this to the lake and the fish
populations (Wright 1955, Fish 1960). These studies
were not followed by any continuing research commitment
and hence, they could not provide the kinds of
information needed even for crisis management.

At about the same time as the Lake Erie studies, a
fishery/limnological study was undertaken on Lake
Michigan. It was more successful in that the objective
was to evaluate fishing gear (Hile 1957).

It was known that the sea lamprey (Petromyzon marinus)
had gotten above Niagara Falls as early as 1921 and had
penetrated the upper lakes in the early 1930s
(Applegate and Moffett 1955). No action was taken
since the impending destruction of the lake trout
stocks by this predator was not recognized and the
Canadian and United States governments soon became
preoccupied with World War II.

The magnitude of the disaster was apparent by the late 1940s as stocks of large fish crashed. By 1950, the lake Huron lake trout population had collapsed and Lake Michigan trout production had decreased by 95%. It was in 1950 that United States Congress finally allocated substantial funds for research on the sea lamprey problem. Funds were used to mount a major program of research on the lamprey and its environments. A major fishery-limnological research program was established and a number of fishery and limnological surveys were undertaken by U.S. Fish and Wildlife Service/U.S. Bureau of Commercial Fisheries on Lakes Michigan, Huron, and Superior in the 1950s. Attention was diverted to Lake Erie in the late 1950s because of further major changes in fish stocks of that lake.

The research on Lake Erie resulted in re-examination of the results of all the previous studies and the astounding conclusion that this lake had undergone major changes, not only in the fish stocks, but in the phytoplankton, zooplankton, and benthic communities, that coincided with long-term increases in the chemical content of the waters (Beeton 1961). These changes were soon recognized as being closely similar to those observed in smaller lakes which had undergone accelerated eutrophication. Studies of the available data for the other Great Lakes showed that all except Lake Superior were undergoing accelerated eutrophication (Beeton 1965). Some symptons of eutrophication, such as blooms of bluegreen algae, were a nuisance, but the major concern was over the changes in the species composition of the fish communities.

Concern over the impact of power plants, especially nuclear plants, led to a large amount of applied research during the 1960s and 1970s to determine the environmental impact of these facilities which use huge volumes of lake water for cooling (Tesar et al. 1985). The major concern was and is over the effect of a thermal rise on the survival of fish. Subsequently, research was diverted to the impact on fish populations of entrainment and impingement of larval fishes.

Many of the changes in the fish populations certainly resulted from sea lamprey predation and probably eutrophication, but many changes likely occurred because of physical changes of habitat, overfishing, and the introductions of exotic species such as the carp, smelt, alewife, white perch, and more recently, several species of salmon.

In any case, concern over the fisheries resource whether for commercial or recreational use has been a major stimulus for much of the research and management of the Great Lakes. In fact, much of our concern about toxic substances results from the knowledge that Great Lakes fishes bioaccumulate contaminants to concentrations which are potentially harmful to humans.

Public health problems associated with the use of the Great Lakes as a water supply and for waste disposal, became another concern around the turn of the century (NRC/RSC 1985). Major attention was given to typhoid, a water-borne disease. Studies were undertaken of the water quality of southern Lake Michigan and the boundary waters of Canada and the United States, especially the St. Marys, St. Clair, Detroit, and Niagara rivers. The treaty establishing the International Joint Commission to deal with boundary water problems was negotiated at this time, in 1909.

Basic Research

The early emphasis on applied problems in response to a series of crises does not imply that basic research was completely lacking. The tempo of basic research has greatly increased recently, building on early subject areas or regional limnological surveys. Before the turn of the century, however, many researchers had to be content with Great Lakes water drawn from the local water intake for their studies.

The first comprehensive study of surface currents in all the Great Lakes, deduced from movements of drift bottles, was published by Harrington (1895). Harrington, former Professor and Director of the University of Michigan Astronomical Observatory, left the University in 1891 to become the first head of the U.S. Weather Bureau. Another University of Michigan Professor, Jacob Reighard, began biological studies of Lake St. Clair at about this time. He proposed to study:

>   ..."the biology of the lakes from the point
>   of view of pure science for the purpose of
>   finding out as far as possible of the facts
>   and making clear as many as possible of the
>   principles"...(Reighard 1899).

In the early and mid 1900s, university laboratories and institutes were established as regional interests in the Great Lakes intensified (Beeton and Chandler 1963). Much of the early work at the Franz Theodore Stone Laboratory of Ohio State University was related to fishery problems. Starting in 1938, however, a full-time staff was recruited to study the general ecology of western Lake Erie. This year-round program, discontinued in 1955, provided a wealth of information on the plankton of the western basin of Lake Erie. A year-round program was resumed in 1973. Two major studies were undertaken by the Great Lakes Research Institute at the University of Michigan, one on Lake Huron in 1954 and the other on Lake Michigan in 1962. These studies involved the first multi-ship synoptic surveys of thermal and chemical characteristics of the lakes. The surveys demonstrated the value of oceanographic techniques applied to lake research, such as the dynamic heights method for determining currents. At the University of Wisconsin, Clifford Mortimer began his studies on the internal waves of Lake Michigan in 1962. These studies were to continue until the present, providing valuable insights into the dynamics of other lakes as well as applications to physical oceanography.

During the 1960s and 1970s, limnologists at the University of Michigan, Ohio State University, and the University of Wisconsin focused on an understanding of primary productivity as one of their central goals. The research demonstrated the basic importance of phosphorus and silica as limiting elements in phytoplankton and diatom growth. Moreover, emphasis on productivity and limiting nutrients laid the basis for increased interest in process-related phenomena, determination of biogeochemical cycling, and food web studies. The research demonstrating the importance of silica as a nutrient influenced by eutrophication laid the basis for some modern controversies.

Interest in the food web grew out of taxomic and fisheries studies of the Great Lakes biota. Detailed studies on phytoplankton taxonomy and physiology, zooplankton vertical migration, and fish population dynamics contributed greatly to an understanding of interactions. The studies on vertical migration demonstrated that the responses to light and temperature by the Great Lakes mysid, Mysis relicta, were similar to those of some marine mysids. Modifications of the species composition of zooplanktonic communities following the alewife introduction to Lake Michigan in the 1960s set the stage for the "top-down/bottom-up" controversy that dominates modern food web symposia.

Unfortunately, the research in response to crises and the scattered results of basic research have not provided a coherent base of knowledge to develop management plans for future use of the lakes.

CONCLUSION

We have two sound reasons for emphasizing the role of research in management of the Laurentian Great Lakes. First, the scientists in the universities, state and federal agencies recognize that the problems which face us today, especially toxic contaminants, cannot be dealt with by crisis management. We need to understand the critical processes of Great Lakes ecosystems in order to manage uses of these lakes in such a way as to minimize adverse impacts. Secondly, research in the Great Lakes can provide us valuable knowledge and insights for research on small lakes and the oceans.

The Great Lakes Water Quality Agreement of 1978 is truly a pioneering international instrument in that it calls for an ecosystem approach to dealing with the problems occurring from use of these lakes (NRC/RSC 1985). In terms of management, however, we must recognize that we can only manage the uses of the lakes since the ecosystems are under the control of nature. The main usefulness of the ecosystem approach is that it provides for an integrative, future-oriented, preventive approach for management.

Most management and research organizations agree to an ecosystem approach for the Great Lakes. Yet we are not taking the steps necessary to implement such an approach - we continue crisis management, especially in dealing with toxic substances. The ecosystem approach means that all the agencies must interact and cooperate in their research. An integrated "use management plan", developed on the basis of our knowledge of how the Great Lakes ecosystems function, should provide for protection and rehabilitation of the quality of the Great Lakes. In terms of toxic substances, such a management plan must deal with questions such as:

How are stresses best measured? and

What are the cost/benefit relationships when all compartments of the ecosystem are taken into account?

The ecosystem approach is not possible at present because we simply have not carried out the coordinated basic research needed to understand how the ecosystem functions. Most of the funds for research have been in support of crisis management. If we are to have an ecosystem approach, then the management agencies must recognize the need to cooperate in setting an agenda for the research which will provide them with the appropriate knowledge. This also means that we need the support of the public and its elected representatives.

Unfortunately, now that we appear to be on the threshold of actually moving to a research agenda which will allow an ecosystem approach, we are faced with serious erosion of funds. The Draconian budget cuts, especially at the federal level, have seriously affected the Great Lakes research community. Federal research laboratories have lost scientific talent and their ability to mount any major new initiatives. Universities have been especially affected since support has dwindled for environmental research and fewer students are pursuing studies leading to a degree in environmental sciences. The Great Lakes states are losing a valuable resource - scientists. The comments which President Harry Truman made in establishing the National Science Foundation in 1950 are certainly worth recalling today:

> ..."No nation can maintain a position of leadership in the world today, unless it develops to the full its scientific and technological resources"...

We should remember that basic research provided the support for over 70 percent of the technological innovations in the United States (NASULGC 1985). The value of basic research is possibly even greater when dealing with environmental problems.

REFERENCES

Applegate, V. C. and J. W. Moffett 1955. The sea lamprey. Sci. Amer. 192:36-41.

Beeton, A. M. 1961. Environmental changes in Lake Erie. Trans. Amer. Fish. Soc. 90:153-159.

Beeton, A. M. 1965. Eutrophication of the St. Lawrence Great Lakes. Limnol. Oceanogr. 10:240-254.

Beeton, A.M. and D. C. Chandler 1963. The St. Lawrence Great Lakes. In: Limnology in North America. ed. D. G. Frey, pp. 535-558. Univ. Wis. Press, Madison, Wisconsin.

Bissell, J. H. 1887. Fish and fish-culture in Michigan. Seventh biennial rept. State Bd. Fish. Comm. pp. 94-108.

Fish, C. J. 1960.   Limnological survey of eastern and central Lake Erie 1928-29.   U.S. Fish and Wildl. Serv., Spec. Sci. Rept., Fish. No. 334, 198 pp.

Harrington, M. W. 1895.   Currents of the Great Lakes, as deduced from the movements of bottle papers during the seasons of 1892, 1893 and 1894.   U.S. Dept. Agr., Weather Bur. Bull.B. 14 pp.

Hile, R. 1957.   U.S. federal fishery research on the Great Lakes through 1956.   U.S. Fish and Wildl. Serv., Spec. Sci. Rept., Fish. No. 226. 46 pp.

National Association of State Universities and Land Grant Colleges 1985.   In the national interest.   Higher education and the Federal Government: the essential partnership.   NASULGC. Washington, D.C. 27pp.

National Research Council/Royal Society of Canada 1985. The Great Lakes Water Quality Agreement: an evolving instrument for ecosystem management.   Nat. Acad. Press. Washington, D.C. 224 pp.

Reighard, J. E. 1899.   A plan for the investigation of the biology of the Great Lakes.   Trans. Amer. Fish. Soc. 28:65-71.

State Board of Fish Commissioners (Michigan) 1887. Seventh biennial report. 130 pp.

Tesar, F, J., D, Einhouse, H. T. Tin, D. L. Bimber and D. J. Jude 1985.   Adult and juvenile fish populations near the D. C. Cook nuclear power plant, southeastern Lake Michigan, during pre-operational (1973-74) and operational (1975-79) years.   Great Lakes Res. Div., Univ. Mich., Spec. Rept. No. 109. 368 pp. +microfiche appendix.

Wright, S. 1955.   Limnological survey of western Lake Erie.   U.S. Fish and Wildl. Serv., Spec. Sci. Rept., Fish. No. 139. 341 pp.

CHAPTER 19

A CITIZENS ORGANIZATION AND ITS ROLE IN PUBLIC POLICIES

Thomas L. Washington

Executive Director, Michigan United Conservation Clubs, Lansing, MI 48909 USA

·This paper summarizes the role that citizens organizations play in designing, developing and implementing public policies that can protect the world's large lakes from the threat of toxic contamination.

Over the past 49 years, through its citizens education programs, lobbying efforts, and in some cases through litigation, the Michigan United Conservation Clubs (MUCC) has spearheaded citizen efforts to ensure the quality of the aquatic ecosystem of the Great Lakes Basin. Chartered in 1937, the MUCC is a grassroots association of citizen conservationists and outdoor sports enthusiasts who are dedicated to the conservation of natural resources and to the development of sound and rational public policies that protect those resources.

In the half-century that MUCC has spoken out for the cause of conservation and environmental protection, we have witnessed the dramatic growth of toxic substance contamination of our Great Lakes resource and of nearby groundwater supplies throughout the region.

History clearly shows that both the public and private sector response to toxic and hazardous waste disposal problems was, at the very best, shameful, and at the very worst, criminal. From the onset of Americas' industrial revolution until the mid-1960s, state and federal officials maintained a "hands-off" attitude toward private business engaged in the manufacture, transportation, use and disposal of toxic materials. For decades there simply was no established regulatory framework in place to safely manage the more than 57 million tons of toxic waste generated annually by industry.

Finally in the 1960s, people across this nation and worldwide began to sense that something was terribly wrong-and they began to act.

The root of their tireless campaign was fear. Fear of the unknown chemical hazards that lurked in their backyards and in their supply of drinking water. Fear of the long-term effects of these toxic materials on their children and grandchildren. Most importantly, fear that industry officials and government regulators were ignoring their complaints and failing to take the proper steps to mitigate the health and environmental hazards.

As individuals fighting powerful corporations and insensitive government officials, their voices were never heard. As members of powerful citizens organizations like MUCC, their campaign grew dramatically in scope and magnitude - becoming one of the most widely publicized news stories in history.

Now, when these large citizen groups speak as one voice-- government and industry listen.

At MUCC, like other citizen organizations, a variety of tactics and methodologies are used to impact the development and implementation of environmental policy.

First and foremost of these methodologies is public education. Working with other environmental groups throughout the state and the nation, MUCC acts as an independent clearinghouse for information on such environmental matters as water quality, conventional pollution, toxic and hazardous substances and a host of other conservation and natural resource topics.

At MUCC we believe that effective citizen involvement must be preceded by citizenry.

Secondly, MUCC acts as a watchdog organization - constantly monitoring the activities of business and government to ensure that laws, rules and other environmental safeguards are strictly enforced. Through the eyes and ears of MUCCs more than 100,000 members in the state of Michigan, MUCC maintains a constant vigil over our precious natural resources.

The ability to impact the development of natural resource management policy is another key factor in the role of a citizens organization. Commonly referred to as "lobbying," this activity involves the education of public policy decision makers on the organizations official position on issues. At MUCC, public policy positions articulated by the membership are communicated directly to government officials.

A strong and active membership, coupled with a keen knowledge of conservation issues, gives MUCC significant influence over the development of natural resource management policy in Michigan.

All too often, when industry officials and government regulators fail to properly address serious environmental concerns, citizens organizations like MUCC must turn to judicial bodies for action. Although expensive and time-consuming, litigation is frequently employed by citizens organizations in seeking swift remedies to potentially irreversible environmental problems. When attempting to halt the threat of serious environmental harm to a resource as critical as the Great Lakes, citizens organizations like MUCC are often left with no viable alternative.

Those, in a very condensed form, are some of the methods employed by citizen organizations throughout America and the world in an effort to halt the threat of toxic substance pollution of the earths large lakes.

As the proliferation of toxic substances make our large fresh-water lakes increasingly vulnerable to poisonous pollutants, it is essential that citizens organizations maintain a constant, worldwide vigil to protect these critical aquatic resources.

Working together in one, united effort, we can ensure that the large lakes of the world will never again be poisoned by callous indifference.

CHAPTER 20

GREAT LAKES UNITED - AN INTERNATIONAL CITIZEN'S
ORGANIZATION FOR ENVIRONMENTAL ACTION

Robert A. Boice

President, Great Lakes United, Watertown, New York, USA

It is satisfying to relate a success story in which the
author has been personally involved for the past four
years.   This success story involves the organization
and mobilization of citizen interest in environmental
action within the Great Lakes - St. Lawrence River
drainage basin.

Exactly four years ago this week an idea was born on
this very island (Mackinac). With initiative supplied
by the Michigan United Conservation Clubs and financial
backing from the Joyce Foundation, a meeting was held
here to explore ways to unite environmental and
conservation efforts throughout the Great Lake Basin.
More than 100 people representing 55 organizations from
the eight States and two Canadian Provinces bordering
the Great Lake system attended. All had one primary
concern - to conserve and protect the natural resources
and environment of the largest body of available fresh
water in the world - the Great Lakes and St. Lawrence
Basin.

The challenges facing the formation of a successful
coalition were many. The main challenge was the size
of the region. From Duluth, Minnesota on the western
tip of Lake Superior to the eastern tip of the Gaspe
Peninsula, Quebec, the waterway covers close to 3,000
kilometers. A straight line between these two points
extends for over 1900 kilometers. The width of the
drainage basin from north to south varies from 150 to
900 kilometers. The vast size of the basin poses a

challenge to both transportation and communication for a volunteer organization. The basin extends to the heartland of both Canada and the United States and hold major industrial, agricultural, tourist and recreation centers of both countries. Some 50 million Americans and Canadians live within the Great Lakes - St. Lawrence River Basin. Although the governments of both countries come from similar backgrounds, there are considerable differences on both the local, regional and national level. Even with these differences, and more, the representatives attending the Mackinac meeting were successful in focusing on a common objective - improving their environment.

Out of the Mackinac meeting came a document which was to be the keystone to the eventual birth of Great Lakes United - a Charter. The Charter provided a bond on which all parties concerned could find common ground to work together. That Charter reads as follows:

GREAT LAKES CHARTER ADOPTED AT THE MACKINAC ISLAND MEETING MAY 22, 1982:

WHEREAS, the Great Lakes are the greatest fresh water system on earth; and

WHEREAS, 50 million people live within and influence the Great Lakes ecosystem and millions more receive economic, recreational and spiritual benefits from them; and

WHEREAS, there is a need for economic strategies compatible with maintenance of the natural system; and

WHEREAS, there is a need for cooperative and coordinated citizen action on behalf of the Great Lakes; and

WHEREAS, we have agreed on the need for such action on the critical issues of:
- Water quality;
- Hazardous and toxic substances;
- Atmospheric depositions;
- Regulation of levels and flow including diversions;
- Fish and wildlife management and habitat protection;
- Energy development & distribution;
- Land quality & land use practices
- Navigation issues such as winter navigations, additional locks, channel modifications, etc.; and
- Public support for Great Lakes ecosystem research, education and management,

THEREFORE, we resolve to establish a Great
Lakes organization to provide an
information exchange and a forum
for working together on these issues.

With the above Charter as a guide, a Charter Committee,
elected at the Mackinac meeting, met during that summer
and early fall to draft organization by-laws, with a
deadline to hold an organization meeting in the fall of
1982. On November 21, 1982 the organization meeting
was held in Windsor, Ontario and the by-laws were
officially ratified. The name GREAT LAKES UNITED was
officially adopted and a Steering Committee was elected
to continue development of the organization structure
and to put together the first annual meeting. On May
6, 7 and 8, 1983 a little less than one year from the
Mackinac meeting, Great Lakes United held its first
annual session in Detroit, Michigan.

Organizations throughout the Great Lakes Basin having
an interest in the coalition were sent an invitation.
The caption on that invitation read:

...."To Unite and Strengthen the Conservation
and Environmental Voice of the Great Lakes -
St. Lawrence Basin"...

Some 70 organizations responded. At this date the
organizational membership of Great Lakes United has
more than doubled. By 1987 we expect over 300
organizational members with close to 2 million
constituents. The membership of Great Lakes United
represents many segments of the basin population. It
is a coalition of conservation, environmental,
sportsmen, union, education, government, small business
and corporate organizations.

The 1983 Detroit annual meeting marked the formal
existence of Great Lakes United. During the next three
years several milestones would be reached by GLU in
organizational development, in education programs, and
involvement in environmental action issues. Briefly,
some of those milestones include:

-Incorporation
-Attainment of 501 c3 Charitable Status
-Publication of a membership newsletter
 "Action Update"
-Task-force development with printed
   environmental position statements
-Success in meeting the winter navigation
   challenge

-"Great Lakes Week" proclamation and
  activities
-Success in obtaining a financial grant to
  hire staff and establish an office
-Hiring of an Executive Director
-Establishment of international headquarters
  at Medaille College at Buffalo, N.Y.
-Development and promotion of Great Lakes
  Week in Washington, DC
-Receipt of a grant award to hire a person to
  develop a membership campaign and produce
  a quarterly organization publication
-Receipt of a grant award to hire a
  coordinator for a Basin-wide public
  hearing program on Great Lakes water
  quality.

No doubt some of the events which took place during the past three years have been missed in this list of milestones. Possibly the most notable is the ongoing commitment and involvement of not only Board members and officers but many, many other volunteers who were instrumental in these development years of Great Lakes United. One of the greatest strengths has been the continual effort to develop and improve a communication system plus maintaining an ongoing dialogue, not only with members and associates but with government leaders at the federal, provincial, state and local levels, including the regional and national agencies and organizations serving the Great Lakes Basin.

For the first year and a half of the existence of Great Lakes United, membership fees and donations from organizational members, particularly those serving on the Board of Directors, kept the organization solvent. A welcome boost came with the award of foundation grants to hire professional and support staff, set up an office and put a communication system in place. Foundations that have contributed financially to the success of Great Lakes United include the Joyce Foundation, Donner Foundation, Public Welfare Foundation, Gund Foundation and the Harder Foundation. One of the most significant highlights was the establishment of the international office of Great Lakes United in the "White House" at Medaille College. The cooperative agreement between Great Lakes United and Medaille College has been instrumental in the organization's growth.

To assess the progress of Great Lakes United - or to put it in a bottom line mode - one must ask:

"What has Great Lakes United contributed that
was not being done prior to its existence?"

To evaluate, one must first review the original Charter
and compare the activities of Great Lakes United and
progress with the six objectives stated in the by-laws,
which include:

-Education
-United States-Canadian cooperation and co-
 ordination
-Promotion of an Ecosystem approach to Great
 Lakes management
-Promotion of environmental/economic
 compatibility
-Development of strong public support.

Most assuredly, the organization has a long way to go
to fully encompass the above goals and although the
authors could certainly be labelled "biased", matching
the brief list of milestones reached over the past
three years, with both the Charter and the by-laws
objectives, there is strong evidence that Great Lakes
United is on course and meeting the challenges set
forth at the Mackinac meeting.   With continued
volunteer commitment, plus the tremendous advantage of
having a competent, energetic and eager professional
and support staff, Great Lakes United is not only
measuring up to the rigid demands and challenges but is
on course to a bright and productive future.

Although a multitude of environmental protection laws,
regulations and policies have been established by the
governments having jurisdiction in the Great Lakes -
St. Lawrence basin, there is still much to be done.
Focusing on environmental problems yet to be conquered,
promoting and conducting education programs for the
general public to increase understanding and awareness
of the complexity of our environment, pushing for
greater financial support, more timely implementation
and stricter compliance of present and future
environmental regulations - for the benefit of both -
our economy and our environment - this is the mission
of Great Lakes United.

CHAPTER 21

INTERNATIONAL CITIZEN ACTION TO PROTECT THE WORLD'S LARGE LAKES

John Jackson

Vice-president, Great Lakes United, 24 Agassiz Circle, Buffalo, New York 14214

ABSTRACT

Thirty-three of the world's large lakes cross international boundaries. In order to protect such lakes, citizens must work together on an international basis. Great Lakes United is a example of a mechanism that encourages international citizen action. Organizing at an international level creates difficulties because of differences in language, political systems, culture and style. But, as the example of Great Lakes United shows, these differences are overcome by a shared desire to protect the large lakes that are an integral and meaningful part of people's lives regardless of which nation they live in.

Thirty-three of the world's large lakes cross international boundaries. Only by cooperative action between or among nations can these lakes be fully protected.

Reaching agreement on a bi-national or multi-national basis on the basic principles that will set the guidelines for action pose a major difficulty in carrying out such cooperative actions. To truly protect the world's large lakes, requires setting lofty goals to strive towards and measure ourselves and

others by.   These principles must be non-exploitive,
recognizing the inherent value of the lakes and the
basic rights of wildlife and the environment.

We need international treaties and agreements asserting
these principles.   The 1978 Great Lakes Water Quality
Agreement between the U.S. and Canada is an example of
such an agreement.   Among its lofty goals are the
"virtual   elimination"   and   "zero   discharge"   of
persistent toxic substances.

Such   agreements   are   meaningless, · however,   without
international   arrangements   to   encourage   movement   by
each country towards these goals and without mechanisms
to monitor progress.

But we need much more than international governmental
mechanisms.   Over and over, it has been proven that the
only way to truly protect our large lakes is by citizen
action.   Those living around the lakes are the ones
most directly affected by them, who share in their use
and enjoyment and value the lakes for their multitude
of essential and delightful facets.

In the cases of large lakes bordering more than one
country, mechanisms need to be developed to encourage
citizen action to occur on an international level.   For
it is only by bringing pressure to bear in a united
way, on all responsible jurisdictions simultaneously,
that there is any hope to protect our lakes.   Too often
we are separated by political institutions, cultural
differences,   language   and   ideology.   But   these
divisions are easier to overcome when our mutual love
of a shared treasure - the large lake is recognized.

Mechanisms are needed to encourage citizens to get to
know each other, to discover and define their shared
goals and to speak loudly, in a non-national way, to
responsible government authorities and industry when
they are being negligent or rapacious.

Great Lakes United is an example of an organization
that encourages citizens to achieve their potential in
fulfilling this kind of role.   The success of Great
Lakes United points out the possibilities for citizens
from different countries to work together to solve
large lakes problems.

The differences between the U.S. and Canadian peoples
are less than between many countries sharing large
lakes; nevertheless, there are important differences
and, therefore, the development of Great Lakes United
is instructive.   The following are a few of those
differences as seen from the perspective of a person

who has been a Canadian member on a Great Lakes
United's board for the past three years:

There have been some language problems, French-English
differences, and even in English-English exchanges of
words that are understood differently in the two
countries. This problem has been relatively minor and
usually mainly a source of humour. It has, however,
limited our ability to develop membership in Quebec,
the recipient of the toxics that spew down the Great
Lakes system, through the St. Lawrence River to the
Atlantic Ocean.

The strikingly different political systems and
government decision-making processes in the two
countries have created more important difficulties.
How often have the Canadian members sat in bewilderment
at board meetings as the U.S. members discussed who is
the most important senator or house representative to
lobby on a particular bill? What are these critical
"conferees" that they talk about? And why don't they
understand why a guide on Canadian legislator's voting
records makes no sense to us? But we are gradually
learning about each others political systems. This
learning process was given a considerable boost for
those of us who spent a week in Washington, D.C. on a
lobby week in October 1985. We plan to give our U.S.
members a similar opportunity to experience the
Canadian political system in the fall of 1986.

A significant difficulty in working together arises
from the juxtaposition of a super-power with a middling
power and the fears that engenders in the citizens of
the lesser power. These political realities extend to
the citizen level as well. We fear that our interests
and our agenda will be submerged by the more aggressive
style of the residents of a superpower.

Overcoming these difficulties takes time, patience, and
above all a willingness, indeed a desire, to learn
about the other and to be sensitive to and accepting of
the needs and differences of the other. It also
requires appreciating and valueing those differences.
Out of this comes the ability to transcend political
boundaries to work together creatively.

Working together internationally at the citizens level
can sometimes affect government actions more
effectively than on a single nation basis for several
reasons. Sometimes the moral suasion of saying it is
our international responsibility to take action has a
special power that a purely indigenous problem may not
have. It adds a novel dimension to the situation which
draws the attention of both the media and politicians.

It helps make the issue non-partisan. Above all, it brings a new, broadened perspective to the issue, often resulting in a deepened understanding and new insights into the problem.

The Citizens Hearings on Great Lakes Water pollution that Great Lakes United will be conducting is a prime example of a mechanism that citizen's groups can use to generate an international voice on large lakes problems. Between July and October of 1986, a Great Lakes United task force will visit 19 Great Lakes centres in Canada and the United States. Hearings will be held in each location to encourage people throughout the basin to articulate their concerns and to propose ways of dealing with the contamination problems in the Great Lakes. Concerns and suggestions will be compiled and presented to the responsible governments on both sides of the Canadian - U.S. border. This project will demonstrate that the public throughout the Great Lakes Basin have transcended narrow national perspectives and expect their governments to do likewise in taking aggressive actions to protect the Lakes.

We must not sit back and hope that our governments will take care of the world's large lakes. We, the people who live around these lakes and delight in them, must be strong advocates on their behalf. When the lakes are international, we must work together internationally to protect them.

CHAPTER 22

INTERNATIONAL AGREEMENTS AND STRATEGIES FOR CONTROLLING
TOXIC CONTAMINANTS

Orie L. Loucks[1] and Henry A. Regier[2]

[1]Holcomb Research Institute, Butler University,
Indianapolis, Indiana 46208
[2]Department of Zoology, University of Toronto, Toronto.

INTRODUCTION

After a century of responses to pollution in the Great
Lakes Basin, "successful" control actions are coming to
be viewed as those which achieve at least partial
remediation of local problems, if not lake-wide or
basin-wide improvement. Successful control has
occurred, but is more difficult to demonstrate where
pollution results from a complex of abuses whose
control involves more than one agency of a state,
province or local government. The forty-two "areas of
concern" around the Great Lakes (IJC 1985) are all
illustrative of these multi-source and multiple
impairment situations. Progress has been particularly
slow in dealing with pollution problems which,
following hydrologic and atmospheric transport, have
expanded in scale to involve an even larger number
jurisdictions, including state, provincial, local and
national governments (NRC/RSC 1985). Numerous
jurisdictions are the rule rather than the exception as
we consider in this paper the problem of controlling
toxic contaminants reaching large lakes in other parts
of the world.

Increasingly, the question of how to protect the
world's large lakes involves controlling pollutants
from hemispheric or global sources, particularly the

persistent pesticides, fluorocarbons, and oxides of sulfur, nitrogen and carbon.   Proposed programs for dealing with multinational pollution problems are still framed, operationally, within regional agreements, usually between two or up to a few countries.   We can still only speculate as to whether toxic control programs involving a larger number of jurisdictions will even be feasible.   The objective of this paper is to review the recent experience around the Great Lakes in North America, and around the Baltic Sea in Europe, to determine what procedures seem to have fostered multi-national protection of large lake ecosystems.

Many examples illustrate that, at the local level, control of specific pollution problems can be achieved within a decade, at least where there is sufficient public concern and cooperation (CEQ 1979).   At the level of regional water quality impairments, as in the Great Lakes where there are twelve governmental jurisdictions, the time lag for response is in decades. The slow progress in identifying the problem and then controlling phosphate enrichment is illustrative of this time frame.   One must ask now whether it is possible to extrapolate to the problems associated with hemispheric or global sources of contaminants and infer even larger lag-times before a multinational community can move toward their resolution.

We propose to address these questions by looking at the processes by which progress was made in two cases of multi-national agreements to mitigate resource degradation.   These are the Great Lakes Water Quality Agreements of 1972 and 1978 between the United States and Canada, and a series of agreements in a similar time-frame that have influenced programs in the nations bordering the Baltic Sea.   Technical and scientific evidence will be considered, as well as institutional arrangements and the apparent strategies of the political and administrative leaders in moving forward (or temporarily avoiding) these agreements.   One clear concern must be to answer why, until this second world large-lakes conference there was no evidence either in the Baltic nor in the Great Lakes Basin that senior administrative or political leadership would adopt strategies sufficient to prevent long-term impoverishment of large lake resources.   The announcement in May 1986, of a Cooperative Agreement for controlling toxics in the Great Lakes announced by the political leadership of the Great Lakes Basin may be a significant step in a new direction.

ELEMENTS OF CONTAMINANT CONTROL STRATEGIES

For either the Great Lakes or the Baltic Sea Basin, we
must consider first the large spatial and temporal
range of inputs and system responses, and the
complexities of the toxic contaminant transport and
effects processes on resources. A straightforward
strategy for analyzing these issues would require
appropriate documentation of several types, including,
but not limited to scientific information on:

    1.The composition of toxic releases at their
    sources, i.e. at point of release of
    commercial products to the market, or of all
    residuals released into the environment;

    2.Interactions among contaminants, as during
    burning or photooxidation after their
    release;

    3.Transport geographically, as with air or
    water movements over regions and across
    boundaries;

    4.Subsequent fluxes between the media of air,
    water and soil;

    5.Impacts on non-human biota; and

    6.Transmission to and cumulative effects on
    humans.

Of these types of information, the first, third and
sixth appear to involve national policies and
responsibilities and are, inherently politically
sensitive, in addition to having complex scientific
elements. Together with the other three types of
documentation needed, each one also scientifically
complex, projection of outcomes become potentially
intractable with conventional reductionist scientific
approaches.

Let us expand on point 1, the question of "sources".
In earlier years, sources were identified either as
industrial or municipal discharges, to be regulated or
controlled by permit procedures designed to reduce the
total discharges. From an ecological perspective,
current sources of contaminants to aquatic environments
are recognized as being much broader, and include the
diffuse loadings from the atmosphere, from sediments,
from groundwater inflow, from surface runoff and from
the conventional point sources. Thus documentation of

"sources" now needs to consider the ultimate sources, where persistent contaminating substances are manufactured or used prior to their release in the environment. Unfortunately, however, such information is often protected because proprietary information about formulations could be inferred from it.

Passive transport of contaminants across jurisdictional boundaries, in the flow of water and air masses, also may lead to inter-governmental charges of neglect of neighboring resources. Dischargers who release wastes freely into air or water masses that subsequently cross boundaries have the advantage of an indirect subsidy not only from citizens in its jurisdiction, but from citizens in other jurisdictions affected by those wastes. To the extent that such waste loadings are authorized by the source jurisdiction, a kind of international "subsidy without representation" may be taking place.

The consequences of long-term exposures of humans to toxic contaminants also is a politically and culturally sensitive issue. Governments are often thought unusually slow to fund data collection and research on the long-term risks from the combinations of chemicals found in the food chain of large-lake systems. Slow progress with the other types of documentation needed for a control strategy may be due to inertia within the ranks of professional scientists. Individuals have been known to make a virtue out of scientific conservatism, despite the evidence of substantial risks explicitly linked to that conservatism.

In addition however, we know there are large-scale linkages among the source, fate and effects processes above. There have been, for example, results presented at this conference on long-distance transport of toxic substances to Lake Superior and to other large lakes around the world remote from any significant industrial discharges. There are many other examples of long-distance transport, including transport between continents, but we are still unable to identify them. Thus, a strategy for managing the effects of toxic substances in remote aquatic environments requires information on intercontinental source/receptor relationships, i.e. a truly global scale.

THE US/CANADA GLWQA

The United States and Canada began addressing the problem of transboundary pollution almost four-score

years ago through the 1909 Boundary Waters Treaty
signed by both countries.   This Treaty established a
bi-national agency, the International Joint Commission
(IJC), with certain constraints designed so that
neither government gave up its respective prerogatives
to determine policy internally, and not have to take
actions it did not wish to take.   Bi-national
arrangements, including the use of a "reference" to set
up a definitive joint study on specific problems, were
provided to fully document the need, the cost, and the
prospective benefits and to find flexible ways of
responding to pollution problems. The special problems
of the Great Lakes, however, were such that further
agreement, the Great Lakes Water Quality Agreement, was
eventually signed in 1972 and expanded in 1978 (NRC/RSC
1985).

The 1978 Agreement states that its purpose is:

> ..."to restore and maintain the chemical,
> physical, and biological integrity of the
> waters of the Great Lakes Basin ecosystem.
> In order to achieve this purpose, the parties
> agree, to make a maximum effort to develop
> programs, practices and technology necessary
> for a better understanding of the Great Lakes
> Basin ecosystem and to eliminate or reduce to
> maximum extent practicable the discharge of
> pollutants into the Great Lakes system".

The extension of the 1978 Agreement to include an
ecosystem view was new for an international agreement
and for such a complex system, and its implications are
still being evaluated and applied. We recently co-
chaired a non-governmental review of its effectiveness
(NRC/RSC 1985), with support from Donner Canadian
Foundation and the William H. Donner Foundation. Our
Committee found "that major progress was achieved
toward meeting the commitments of the 1972 Agreement,
and that progress with respect to the 1978 Agreement
goals may become evident soon." These agreements have
come to involve a number of agencies in each of twelve
jurisdictions: two federal, two provinces and eight
states. Regional and municipal levels of government
also are increasingly involved.   Higher levels of
international oversight, such as the International
Court at The Hague and UN agencies, have not yet become
involved in Great Lakes issues.

Two important joint institutions were established
through the Agreement: the Water Quality Board and the
Science Advisory Board.   The main responsibilities of

these two joint institutions can be summarized as follows:

1.**Data Collection, Analysis and Distribution**: The joint institutions do not collect data in their own right, but rely on measurements obtained by agencies of the various governments involved (state, provincial and federal), and from certain university and contracted studies. The data are brought together on a municipality, lake or region-wide basis by the joint institutions for publication and distribution.

2.**Advice and Recommendations by the Joint Institutions**: To implement the responsibility for providing advice to the two national governments, the IJC relies on the joint institutions established in the Agreement, the Water Quality Board, the Science Advisory Board, and the Regional Office. The IJC has been generally successful in providing advice through these means. The Biennial reports prepared under the Agreement by the IJC contain numerous specific program assessments and recommendations for action by the governments of Canada and the United States.

3.**Assistance in the Coordination of Joint Activities**: The Water Quality Board, under its terms of reference, has the particular responsibility on behalf of the IJC to ..."undertake liaison and coordination between the institutions established under this Agreement and other institutions and jurisdictions which may address concerns relevant to the Great Lakes basin ecosystem so as to ensure a comprehensive and coordinated approach to planning and to the resolution of problems, both current and anticipated."...The Science Advisory Board's terms of reference gives a more limited role of advising the jurisdictions of relevant research needs, soliciting their involvement and promoting coordination.

4.**Investigations**: Under the Agreement, the IJC is to investigate subjects related to the Great Lakes basin ecosystem, ..."as the parties may from time to time refer to it." In addition, the Water Quality Board and the Science Advisory Board may conduct investigations pursuant to their general terms of reference. However, since the

signing of the 1978 Agreement, the parties
have not asked the IJC to conduct any
investigations. Rather, all investigative
activities have been initiated by the Water
Quality Board and the Science Advisory Board.

5.**Public Information:** The 1978 Agreement
assigns responsibility for "public
information service" for the programs of the
International Joint Commission, the Water
Quality Board and Science Advisory Board to
the Regional Office. Both the 1982 and 1984
biennial reports of the IJC have stressed the
value of public information and education in
building support for the Agreement's goals.
Still, limited resources have been committed
to public information, and results have been
irregular (NRC/RSC 1985).

In viewing the progress to date under the Agreement and
through the joint institutions, the NRC/RSC Committee
(1985) made the following major findings:

1.The past century can be shown to have had a
record of resource degradation in the basin,
intruding more and more deeply into
ecosystemic processes, expanding in area from
local effects, to whole-lake effects, to the
whole basin, and extending in duration of the
impact from brief to almost irreversible;

2.The causes of these impairments are now
more complex and have become less evident;
and

3.The sequence of degradation, subsequent
corrective measures, which are only partially
effective, and the prospects of new
environmental impacts have not yet been
brought under control.

Looking to the future, the Committee also found that
there is now a large set of issues that could trigger
even greater requirements for farsighted management
measures, at an even larger scale in the future. For
example, existing capital infrastructures established
for past corrective actions will age and become
ineffective. Independently, consumptive use of water
within or outside the basin is likely to increase
demand for water and increase overall diversions from
the ordinary flow pattern. Finally, there is also the
prospect of large-scale and very long-term climate
change that will affect the functioning of the system.

The process that led to the 1978 Agreement needs to be extended and strengthened so as to deal proactively with forthcoming issues rather than only reactively to past mistakes.   This will require more interactive governance involving collaboration with other bi-national arrangements for Great Lakes fish, North American migratory birds, Great Lakes water levels and flows, etc..

ECOSYSTEMIC ISSUES IN THE BALTIC SEA

The actual processes of international governance in the Baltic at the present time appear not to be very different from those present in the Great Lakes region, but how they developed is quite different. The inter-jurisdictional governance in the Great Lakes emerged gradually in an ad hoc, piecemeal fashion. The political boundaries were established in the middle of the lakes and rivers and the entire set of waters had been demilitarized early in the 19th century. Water levels and flows came to be managed bi-nationally early in the 20th century, while fisheries came to be managed as a program of rehabilitation beginning in the 1950s. Water quality issues were not addressed effectively on a multi-jurisdictional basis until the 1970s. Air quality issues still are not generally under any bi-national management. The prospects for climate change and significant changes in Great Lakes hydrology has not yet appeared on a bi-national agenda in this region.

In the Baltic Sea area, two conventions were signed in the mid-1970s:

> 1.the Gdansk Convention on Baltic Sea Fisheries and
>
> 2.the Helsinki Convention on the Baltic Environment.

Three other broader, international initiatives agreed to during the early 1970s paved the way for these conventions:

> 1.the Inter- governmental Maritime Organization's action concerning pollution from shipping,
>
> 2.the principles and action plan from the Stockholm Conference on the Human Environment, and

3.the de factor consensus that emerged from
the Law of the Sea negotiations on extension
of national control out to 200 miles in the
sea.

The latter led to "assignments" (not yet fully
resolved) of the entire Baltic with respect to various
uses.   One must remember that the Baltic is not
demilitarized.

A   fourth   international   initiative   of   historic
importance, taken by Baltic and other western nations,
also   served   as   a   precursor   to   the   two   Baltic
conventions.   Early in this century the International
Council   for   the   Exploration   of   the   Sea,   ICES,   was
created and ultimately headquartered at Charlottenlund,
Denmark.   This council has long had responsibility for
functions similar to that of a "standing reference" (as
used   under   the   U.S./Canada   International   Joint
Commission),   along   with   a   number   of   different
international commissions of the North Atlantic region.
ICES   serves   this   role   for   fisheries   as   well   as   for
environmental   issues,   on   a   kind   of   contractual
arrangement with the parties to the Baltic conventions.
ICES   has   a   tradition   of   maintaining   scientific
impartiality   in   political   matters,   a   policy   that   is
reinforced   by   moral   support   from   the   non-governmental
International   Council   of   Scientific   Unions   in   Paris,
the   Food   and   Agriculture   Organization   of   the   United
Nations, and other international agencies.

Thus,   in   the   Baltic,   the   fact-finding   function   is
organized   through   broader   international   conventions
than   are   the   management   of   fisheries   and   water   quality
(which   come   under   the   Gdansk   and   Helsinki   Conventions).
As   in   the   Great   Lakes   Basin,   individual   experts   serve
their   agencies   in   their   own   nation,   but   when   they
participate   in   the   Baltic   Commissions   and   meet   with
experts   from   other   convention   nations   under   the   aegis
of   ICES,   they   serve   as   non-partisan   scientists   to
develop consensus on scientific issues.

In   the   Great   Lakes,   the   IJC   or   through   it,   the   Water
Quality   Board,   exercises   both   fact-finding   and   decision
functions   with   respect   to   some   water   quantity   and
quality   issues;   the   Great   Lakes   Fishery   Commission
(GLFC)   similarly   exercises   both   with   respect   to   sea
lamprey   control   and   many   aspects   of   fishery   management
coordination.   In   the   Baltic,   practical   progress   has
been   registered   with   respect   to   joint   management   of
certain   fisheries   stocks   under   the   Gdansk   Convention.
Less   progress   is   apparent   under   the   Helsinki
Convention,   except   with   respect   to   pollution   from
shipping   and   emergency   measures   related   to   vessel

accidents.   Since   the   Baltic   conventions   have
functioned for only about a decade, and significant
time-lags   exist   for   responses   to   all   large-scale
measures for remedial action, the Baltic Sea example
may become just as effective as the GLWQA.

SCIENCE AS A COMPONENT OF INTERNATIONAL PROTECTION
STRATEGIES

Inter-jurisdictional collaboration in science appears
to have been similar in process if not in structure in
both the Baltic and Great Lakes examples.   In both
cases, data collection and analysis from two or more
jurisdictions involves some form of "reference" with
the expectation that the researchers who participate
will not bias the process in favor of national self-
interest.   However, one expects, and we have already
seen, that certain national interests may be perceived
to be threatened by the scientific findings, or by
their   interpretation   in   the   form   of   recommended
actions.   Thus, each party to these multi-national
agreements has reserved for its own determination
almost all issues of the three types noted earlier that
could be viewed as a matter of national policy, and
therefore politically sensitivity.   The balancing of
these   technical   and   political   issues   in   the
implementation of agreements, or even in joint studies,
is evident in the various papers documenting outputs
from the agreements.   The following seven principles or
inferences   have   been   developed   from   the   authors
experience and with the advice and assistance of
Professor A. P. Grima of the University of Toronto.
They are offered as background to the consideration of
multinational protection strategies for large lakes:

>   1.Weaker   jurisdictions   suffering   from
>   inequitable   practices   by   stronger
>   jurisdictions may be most interested in
>   "fact-finding " in order to arrive at
>   decisions that are "principled" with respect
>   to equity.   Stronger jurisdictions, and/or
>   those responsible for bad practices, may
>   agree to fact-finding on the implicit
>   condition that the terms of reference of the
>   fact-finding process should not relate
>   directly to the broader principles of equity.
>
>   2.Fact-finding tends to focus, at least
>   initially, on biophysical phenomena that may
>   be far removed from the socio-economic,
>   equity and political issues actually under

contention.   These biophysical phenomena may
become implicit surrogates of the political
issues.   The biophysical facts may serve as
the   critical   inputs   to   diplomatic
negotiations, but in what becomes an indirect
and somewhat convoluted rhetoric.

3.Social   scientists   (not   including
economists) and philosophers of ethics have a
tradition of wanting to focus on phenomena
that are more immediately relevant to the
"real   concerns".   This   directness   may
actually   blur   the   distinction   between
objective fact finding (however tenuous the
connection to the issues of concern) and the
political   negotiations   involving   these
concerns, some of which the negotiator must
keep confidential.

4.Fact-finding as part of a reference, is
viewed as essentially a collegial process
among peers (experts) who are non-political
representatives   of   each   jurisdiction.
Relative standing in this collegium is partly
a function of degree of expertise, which can
have a correlation with the relative power
(as   conventionally   perceived)   of   the
jurisdiction represented by the expert.

5.A prerequisite for full participation in
the fact-finding process is the provision of
appropriate   information   by   the   expert's
jurisdiction.   The better the information
and   the   greater   the   expertise   from   a
particular jurisdiction, the more powerful
will be that expert's role in the fact-
finding process.

6.Fact-finding   or   reference   studies   are
typically a slow process.   Several years may
pass before study objectives are defined and
appropriate expert groups convened.   It may
take several years for a group of peers to
develop a collegial process.   The more
"controversial" the issue, in the view of the
more powerful jurisdictions, the slower the
process.

7.As   a   collegium   develops   and   inter-
jurisdictional governance evolves (often over
decades), details of the issue referred to
the fact-finders approach more closely the
real   issues   under   contention.   The
..."asymptotic   process   of   progressive

approximations"...relates to spatial proximity, temporal proximity, causal proximity and political proximity of the participants to the centres and personalities of power.

These inferences highlight the dichotomy between the university community in considering toxic pollutant protection strategies, and the government officials who must make and defend the political decisions regarding contaminant control. Government researchers and managers often have difficulty relating to conventions in the non-governmental community and to interdisciplinary research and review activities. The need to transcend the bounds of their civil servant's management responsibility maybe asking too much. On the other hand, these individuals may simply recognize more fully than scientists the political consequences of the broad view of pollutant consequences that is rarely shared by the general public.

CONCLUDING COMMENTS

As yet, there is little that can be recognized as international strategy for controlling contaminants, in a practical sense. However, there appears to be an implicit strategy of long standing on the use of scientific analyses in contentious international issues related to fisheries and to aquatic environments. The common variations of "study reference" appear to be slow and ponderous, perhaps necessarily so for the development of collegial relationships and consensus. An important development at the level of strategy by the political leadership is the announcement of a Cooperative Agreement in controlling Toxics in the Great Lakes by the Governors of the Lake States and Provincial Premiers in May 1986.

Many countries of the world, including Canada and the U.S.A., are now fostering productive creativity based on technical and scientific advances. With respect to what has gone before in recent geological history, uniquely new phenomena are being created in the fields of radiation physics, synthetic chemistry and biotechnology, most with some potential to alter natural environments. The effects on nature and humanity of such new phenomena are empirically unpredictable. A new domain of ignorance is created de novo with each new phenomenon. A toxic substances control strategy is not just a matter of uncovering the existence of ignorance about the risks from these innovations, rather these areas of ignorance did not exist prior to their introduction.

Canada and the U.S.A. have both signed and endorsed various international and inter-jurisdictional agreements, some of which are world-wide in scope and others are more strictly bi-national. Altogether, these undertakings imply a high degree of consensus between these nations on matters related to the control of toxic substances over large areas, particularly in large lakes. Perhaps it is time that a more comprehensive and ecosystemic set of principles and commitments be recognized in such multinational agreements. With these in place, interdisciplinary and inter-jurisdictional research and management strategies could be developed more fully than at present, perhaps sufficient to make better use of scientific understanding of the system and the goals of the resource managers.

Incremental improvements in managing and evaluating the creation of new products, and scientific research on their risks in the environment, will not help in shortening the response time for reactive control measures. Neither will these steps appropriately localize the consequences of mobile, persistent contaminants in the expectation that local governments can act. More explicit progress seems likely by recognizing the two dimensions of the decision process, political as well as scientific, and in improving the institutions by which we bring both together leading up to the decision process.

REFERENCES

CEQ 1979. Environmental Quality - The Tenth Annual Report of the Council on Environmentally Quality. United States Printing Office. Washington, D.C.

IJC 1985. Water Quality Board. 1985 Report on Great Lakes Water Quality. International Joint Commission, Windsor, Ontario.

NRC/RSC 1985. The Great Lakes Water Quality Agreement: An evolving instrument for ecosystem management. National Academy Press. Washington, D.C.,

CHAPTER 23

ECOSYSTEM MANAGEMENT: OVERCOMING JURISDICTIONAL
DIVERSITY THROUGH LAW REFORM

Paul Muldoon

Canadian Environmental Law Research Foundation, 243
Queen Street West, Toronto, Ontario M5V 1Z4

INTRODUCTION

Where an ecosystem encompasses more than one political
jurisdiction, problems arise in formulating,
implementing and managing a coordinated regulatory
framework for environmental protection. Because of the
existence of two national governments and a myriad of
political subdivisions, the Great Lakes basin provides
a classic case of the problems which may occur when
attempting to employ an integrated ecosystem management
approach with a diversity of jurisdictions. In recent
years, the Canadian Environmental Law Research
Foundation (CELRF) has initiated a number of studies,
seminars and publications in an attempt to facilitate
discussion on how to best overcome these problems.

The purpose of this paper is to discuss the findings of
a recently completed CELRF study (Muldoon 1986) which
looked at whether cross-border litigation could play a
meaningful role in overcoming some diversity problems
within the Great Lakes Basin. While the study focused
upon the Great Lakes, the recommendations could be
transposed to other ecosystems composed of a diversity
of jurisdictions. Before the cross-border study
findings are examined, it is worthwhile to first
explore the problem of jurisdictional diversity, and
then some of the previous efforts to cope with the
problem.

THE PROBLEM OF JURISDICTIONAL DIVERSITY

Natural resource management and environmental protection strategies face many challenges. The most often cited one is their low priority on the political agenda. Many in society still regard environmental protection and economic prosperity as contradictory, rather than compatible goals. As a result, environmental protection goals are still perceived as a luxury, as opposed to a prerequisite to the attainment of long-term, sustainable economic growth. Apart from the lack of political will, there is still a need for more scientific research into existing and potential environmental problems, and the development of new perspectives, methodologies and technologies to address those problems. It is necessary to develop and formulate multi-disciplinary, action-oriented programs for ecosystem management.

Another challenge occurs where there are different political units, whether countries, states or provinces, attempting to govern, in whole or in part, one ecological unit, an ecosystem. This problem, often labelled as "jurisdictional diversity", is apparent at the local, regional and global levels. Indeed, diversity issues within a large lakes ecosystem for example, can be considered a microcosm of the problems of environmental management of the global ecosystem. Attempting to find a consensus among the broader international community on what action should be taken to address world-wide concerns (such as carbon dioxide build-up or stratospheric ozone depletion), is very much the same as determining the appropriate course of action for specific problems (such as concentrations of toxic chemicals), among a relatively few number of actors within a regional ecosystem.

The Great Lakes Basin ecosystem provides an excellent example where the study of the diversity problem on a regional level provides a model for global thinking. There are twelve jurisdictions which comprise the ecosystem - two national governments, two Canadian provinces, eight U.S. states, in addition to a large number of county, municipal and other political subdivisions within each. Because each of these jurisdictions share to varying degrees the management responsibility for the Great Lakes Basin, environmental protection strategies, where they do exist, tend to lack any inter-jurisdictional coordination, and as such, appear to be fragmented, duplicitous and ineffectual.

A number of organizations have recognized the problems stemming from a diversity of jurisdictions within the Basin. The International Joint Commission has identified the dilemma on a number of occasions and succinctly described it as follows in regard to the issue of toxic contamination:

The underlying problem....is the absence of an overall Great Lakes Ecosystem strategy for toxic substances control activities that are being carried out under the various pieces of legislation among the jurisdictions. Programs have been compartmentalized under each legislative mandate, and the resources have been allocated accordingly..... This fragmentation has resulted in duplicated activities in some cases, incomplete program coverage in others, and a limited management capacity to effectively address emerging complex problems (IJC 1982, pp. 5-6).

Representatives from a broad range of agencies and groups involved in Great Lakes issues throughout the region echoed these concerns at a workshop (CELRF 1985). Five consequences of jurisdictional diversity were identified:

1.fragmentation of administrative, research, and remedial programs,

2.piece-meal regulatory schemes, thereby leaving legislative gaps over certain activities or certain pollutants,

3.disparity and inconsistency of environmental standards and methodologies employed for establishing those standards,

4.varying levels of commitment to enforce existing laws, and

5.a variety of legal barriers for residents of one ecosystem jurisdiction to participate in the environmental decision-making processes of other basin jurisdictions.

In the end, a diversity of jurisdictions within an ecosystem can impede the goal of integrated ecosystem management. Since the curative actions of one jurisdiction tend to be mitigated by the inaction or contrary actions of other jurisdictions, cooperative, multi-dimensional approaches are needed to prevent further deterioration of the Great Lakes Basin.

OVERCOMING THE DIVERSITY PROBLEM: THE NEED FOR A MULTI-FACETED APPROACH

Many attempts have been made over the years to cope with the problems that may arise from the divided responsibility over the Great Lakes. The channels of diplomatic and bureaucratic intercourse, the formal and informal communications between federal, provincial and state authorities, together with the everyday interactions of administrative agencies, continue to manage the bulk of inter-governmental basin relations. These mechanisms however, tend to be more reactive to specific crises as opposed to being proactive in implementing coordinated management strategies.

Commencing in 1909 with the Boundary Water Treaty and later with the Great Lakes Water Quality Agreements of 1972 and 1978 (IJC 1972, 1978) a bilateral regulatory framework has been developed which creates a set of common principles governing the use of the Lakes and mutual goals directed toward limited discharges of certain pollutants, in addition to other provisions. Too often however, jurisdictions have not implemented laws which are sufficiently stringent to meet the standards established under such agreements. Where adequate standards do exist, there has been a hesitancy to expend the necessary financial and other resources to implement and more importantly, enforce those laws.

The International Joint Commission (IJC), a bilateral commission established under the Boundary Waters Treaty (1909) was an early attempt to overcome diversity problems by establishing a single institution to address mutual concerns about boundary waters, including the Great Lakes (other than Lake Michigan). In early years, the Commission primarily concerned itself with approving applications concerning the use or diversion of boundary waters pursuant to its quasi-judicial powers under the Treaty. In more recent years however, it has emphasized its investigative role in identifying and recommending solutions to transboundary environmental problems, and in particular those relating to the Great Lakes.

It is commonly agreed, in light of the present political realities, that the IJC's powers will not be expanded to give the Commission a more direct role in implementing its recommendations or as acting as an environmental "watchdog" to enforce bilateral pollution standards. Given the structure and history of the IJC, it may be unwise to advocate major reforms to it, since such reforms might impede its ability to fulfil its present important mandate (Munton 1981).

While the national governments were signatories of these agreements, sub-national governments have also put forth some important cooperative regulatory initiatives.   One is the Great Lakes Compact, an agreement among all U.S. states bordering the Great Lakes, which strives toward presenting a coordinated voice for a broad range of issues affecting the states. Another, and perhaps the best known example, is the Great Lakes Charter, a statement of principles agreed to by all state and provincial jurisdictions of the basin to jointly deal with the issues of inter-basin water transfers and consumptive uses.

Apart from governmental efforts, a number of organizations, such as the Center for the Great Lakes and Great Lakes United, have attempted to overcome the diversity problem by conducting research and lobbying on behalf of all interests within the ecosystem. Great Lakes United for instance, was formed a few years ago to present a concerted voice for over 100 of its member organizations from all regions of the Great Lakes and representing many perspectives, including environmentalists, labour representatives, naturalists, scientists and property owners, among many others. Certainly such efforts are useful to instill the conception of the Basin as one ecological community, a constituency unto itself, irrespective of political borders.

To overcome the diversity problem, it is imperative to further develop and refine these traditional approaches to basic problems.   It is also crucial though, to explore alternative mechanisms to build upon previous initiatives.   The CELRF study on cross-border litigation (Muldoon 1986) is an attempt to revitalize a traditional environmentalist tool and to apply it creatively to the contemporary situation.

THE CROSS-BORDER LITIGATION STUDY

The cross-border litigation study was initiated in 1982 to explore what legal barriers were present for individuals or organizations to cross a border within the Basin and either sue a polluter or participate in an environmental decision-making process of another basin jurisdiction.   For example, if residents from Windsor, Ontario could establish that their property interests were suffering damages and their health is at risk due to air emissions from factories located in the State of Michigan, could those residents hire a Michigan lawyer and sue those factories in the U.S.?

If a Canadian factory was dumping mercury in Lake Erie to the detriment of interests in Ohio, could the state then ask its own court to order the Canadian polluter to halt its deleterious activities?   If a hazardous waste disposal company in New York State was intending to build a pipeline to discharge treated wastewater into the Niagara River, could downstream Canadian interests participate in the environmental assessment and permit hearings to the same extent as potentially affected U.S. interests?

While commencing the study, it was assumed that if individuals and groups could become involved in other legal systems within the basin, the public would have a tool to directly address transboundary problems and thus overcome, or at least minimize, the traditional problems associated with a diversity of jurisdictions. In effect, the public would have another means to protect the ecosystem, other than simply voicing their concerns to governmental representatives who are slow to become involved in extra-jurisdictional matters.

The study proceeded through three research stages. The first stage involved identifying the practical role and utility of environmental litigation generally with particular reference to cross-border litigation.   The second stage was a detailed review of all the environmental laws of all twelve jurisdictions bordering the Great Lakes to determine what litigious rights existed and what barriers were present to prevent non-residents from exercising those rights. The final task was to propose some concrete reforms to redress the identified barriers.

THE NATURE AND ROLE OF ENVIRONMENTAL LITIGATION

Litigation provides a mechanism to resolve disputes between competing interests in society.   These interests may be of a private nature, involving disputes between two or more parties, or of a public nature, involving disputes between individuals and the government as to how a regulatory scheme is to be applied or administered (Jaffe 1968, Chayes 1976). In this context, litigation involves more than court actions. Governments have delegated the task of administering and implementing legislative schemes to a variety of tribunals, boards, and agencies.   These administrative creatures are especially prevalent in the realm of environmental regulation. Since they are designed to determine both private and public rights, they serve an important function in the protection and realization of private and public interests, often in a manner similar to a court. For this reason, the term "litigation" includes proceedings before these administrative tribunals, boards, and agencies.

Environmental litigation has several functions (Sax 1970). For example, it can be used to:

>   1.compensate persons suffering environmental injury and prevent further harm through seeking court orders to halt further deleterious activities;
>
>   2.mobilize individuals, community groups and even governments to address certain issues and to provide a forum for the voicing of opinions concerning appropriate solutions; and
>
>   3.act as a "lever" for greater access to governmental decision-making processes in those instances where individuals feel there is no other way to convince complex bureaucracies to listen to their concerns.

In addition to these functions, cross-border litigation imports other attributes because of its inter-jurisdictional character. For instance, it can serve to mobilize individuals, groups and organizations in order that they may participate directly in the resolution of problems which are difficult to solve because of the diversity of jurisdictions. Where governments have failed to rectify environmentally harmful activities primarily affecting other jurisdictions, cross-border litigation allows citizens to transcend political borders and require courts or tribunals to take appropriate measures to ensure the protection of an ecosystem.

Where persons within the ecosystem have the opportunity to cross borders to sue or participate in legal proceedings anywhere within the ecosystem, litigation also serves a community-building function. It allows courts, administrative agencies and other decision-makers within an ecosystem to view all jurisdictions within the ecological community as a whole, as well as providing the public, as members of that community an opportunity to contribute to the well-being and conservation of that community. Indeed, where cross-border litigation is a viable option to protect the Great Lakes, (in the sense that any resident of the ecosystem can use the legal system of another basin jurisdiction), it can better be described as ecosystem litigation.

When reviewing the laws of all twelve jurisdictions of the Basin, it became clear that, for the most part, individuals and groups of one jurisdiction could use the courts and tribunals of other Basin jurisdictions to protect the ecosystem. More specifically, it was found that in seven out of ten instances, there would be no legal barriers to cross-border litigation within the ecosystem. However, in the other instances the border that inhibits a coordinated regulatory framework also, creates barriers to inter-jurisdictional participation in legal proceedings. For the most part these barriers which prevent cross-border litigation from being transformed into ecosystem litigation, are grounded in legal rules that prevent pollution victims of one jurisdiction from exercising their rights to sue for damages or involve themselves in the environmental decision-making processes in the polluter's jurisdiction. Essentially, two general categories of barriers have been found: "common law" and "statutory" barriers.

COMMON LAW AND STATUTORY BARRIERS TO LITIGATION

Common law is binding, yet unwritten law as declared and interpreted by the courts. As the common law developed historically, it seldom allowed courts to adjudicate upon suits with foreign claimants or where elements of the cause of action occurred outside its territorial limits. For example, according to the antiquated "local action" rule, a court could not hear a claim if the injury complained of pertained to land located in another jurisdiction. The local action rule still exists in all Canadian jurisdictions and the state of Indiana. The effect of this rule, for example, for Michigan residents suffering pollution damage to their lands because of actions in Ontario, would be that they could not sue the Canadian polluters in Ontario courts. Instead, the American pollution victims would be forced to sue the Canadian polluters in U.S. courts.

However, pollution victims have always had trouble suing foreign polluters in their own courts because of the narrow rules governing personal jurisdiction. Historically, courts had no power over persons located outside the courts' territorial boundaries. Over the years, exceptions were created in accordance with local rules of the court. However, some jurisdictions such as in New York and Quebec, have made these exceptions extremely narrow. For example, New York courts will only subject a foreign polluter to its process if that

polluter "carries on business" within the state.   If
the victim's court will not assert jurisdiction over
the foreign polluter, and the local action rule
prevents a suit in the polluter's court, the victim is
left without a remedy only because there is a border
juxtaposed between the environmental injury and the
polluting source.

In addition to jurisdictional rules, choice of law
rules may also pose barriers to litigation.   Under
these rules, where a person commences an action in the
polluter's jurisdiction, the court may decide to apply
the law where the pollution victim resides, rather than
its own law.   For instance, if pollution victims from
Pennsylvania sue polluters located in Ohio, the Ohio
courts may determine, in accordance with their choice
of law rules, to apply Pennsylvania law.   A problem
arises however if Ohio law is far more favourable to
the victims than their own law, such as the instance
where the polluter would be exculpated under
Pennsylvania law but held responsible under Ohio law.
In this example, the Ohio polluter would in effect have
a license to pollute so long as the effects are felt
out-of-state.

Many environmental rights are now defined by statute.
These environmental statutes have proved to be a
formidable weapon for individuals to defend the
environment.   Unfortunately, like the common law, many
of these statutes create barriers to cross-border
litigation. Two kinds of statutory impediments are most
pronounced: residency and territorial barriers.   An
example of a residency barrier is found in Indiana's
Standing To Sue Law where the right to sue for
pollution injury is limited to "citizens of the state".
Hence, persons in Illinois suffering harm from Indiana
polluters could not take advantage of this important
law. Residency barriers are also in place in other
circumstances, such as in Minnesota, where only
residents can intervene in administrative proceedings
affecting the environment or petition a court to review
a decision of a tribunal.   In Quebec, persons from
outside the province are excluded from the right to
enforce Quebec's environmental law, even if that
violation is adversely affecting an out-of-province
interest.

If a statute does not have a residency limitation, it
may nevertheless be territorially limited.   Territorial
impediments arise where the environmental statute has a
limited scope or application in that it is designed to
only protect the "environment of the state" (or
province or country), and not the environment
generally.   A tribunal conducting an environmental

assessment would be unable to consider the impacts of its decision on neighbouring jurisdictions or on the ecosystem as a whole. With the exception of the U.S. federal laws and the state of Michigan, every jurisdiction within the Great Lakes Basin has at least one (and often more) environmental statutes that have residency or territorial limitations.

Statutory limitations may give rise to some anomalous situations. For example, Ontario administrative tribunals have an open door policy allowing non-residents to fully participate in the province's environmental decision-making processes. Yet, because Ontario's environmental protection laws specifically limit their application to the protection of the "environment of Ontario", the tribunal is precluded from taking into account extra-provincial interests. In short, non-residents are free to appear before Ontario's administrative tribunals even though those tribunals are not empowered to consider the extra-provincial impacts of its decision. The right to appear, therefore, is an empty right.

EARLIER ATTEMPTS TO OVERCOME BARRIERS

Efforts to overcome barriers to cross-border environmental litigation are not new. Owing in large part to the work of the Organization for Economic Cooperation and Development (OECD), these efforts are often grouped under the concepts of "non-discrimination" and "equal access" (OECD 1974, OECD 1977). The principle of non-discrimination requires a jurisdiction to give the same consideration to the impact its pollution may have on other jurisdictions as it gives to the impact of pollution within its borders. The equal right of access concept is a procedural means of giving effect to the substantive principle of non-discrimination. It requires that a country afford foreign pollution complainants the same rights of access to its legal and administrative systems as it gives to its own citizens. The principles of equal access and non-discrimination were implemented when Sweden, Norway, Denmark and Finland concluded the Nordic Convention in 1974 (Nordic Environment 1974). According to the Convention, individuals who are or may be adversely affected by "environmentally harmful activities" of a treaty state are guaranteed two basic rights: the right to access to the government agencies and domestic courts of the offending state, and the right to non-discrimatory treatment throughout the adjudicatory process.

Two approaches for implementing the equal access concept have been suggested in North America. In 1979, the American and Canadian Bar Associations (ABA and CBA, respectively) approved and recommended to the U.S. and Canadian governments a Draft Treaty on a regime of Equal Access and Remedy in Cases of Transfrontier Pollution (ABA/CBA 1979). This initiative was well-received, even though it has yet to be accepted by the two national governments. In 1982, the American and Canadian Uniform Law Conferences further pursued the topic. These organizations, which attempt to harmonize state and provincial legal systems by drafting prototype laws for uniform adoption, approved and recommended for enactment the Uniform Transboundary Pollution Reciprocal Access Act (1983). While the contents of the proposed Act substantially parallel the ABA-CBA Draft Treaty, the Uniform Law Conferences felt that the rules could be changed more effectively and more expeditiously through the enactment of uniform provincial and state laws than through a bilateral treaty.

To date the Province of Ontario has been the only Great Lakes jurisdiction to enact the uniform law (while four states and two provinces outside of the basin have enacted it.) Although the Act is a positive step to achieve equal access, it is a very modest one (Muldoon 1984). It only applies to court actions, to the exclusion of other proceedings such as judicial review actions and public hearings on environmental assessment, permit applications and proposed air and water quality standards. Further, the Act is reciprocal in nature; the rights it grants are only operative to non-residents from a jurisdiction that has enacted the Act. Because Ontario is the only basin jurisdiction with the Act, no person within the ecosystem can take advantage of the rights it grants. In light of the limitations with previous equal rights efforts, new initiatives are therefore necessary to guarantee equal access to all environmental decision-making forums within the Great Lakes Basin.

An equal access regime in the Basin must be a first step effort in empowering the public to overcome the diversity problem; the second step would be to give every person within the Basin a common set of fundamental environmental rights which could be described as "ecosystem rights".

When pollution victims have "equal rights" to litigate, they can initiate a lawsuit or participate before an administrative hearing anywhere within the ecosystem and stand in the same position as residents of that jurisdiction. However, equal rights only gives the

non-resident litigants the same rights as those vested
in residents of the jurisdiction.   As one commentator
noted, "...if he has few, you have no more..."(Chester
1982, p.86).   Within the Great Lakes ecosystem, this
disparity of rights is quite apparent.   Ontario
illustrates this in an extreme way.   There, the public
is almost totally excluded from important parts of the
environmental decision-making process, such as the
enactment of water and air pollution standards and
permitting processes.   Moreover, the right to bring a
lawsuit for damages or seek a court order to halt a
polluting activity can only be sought by those who
personally suffer an injury over and above that of the
general public.   The government, therefore, is vested
with the sole authority to sue for injury to public
waters (such as the Great Lakes) or pollution affecting
a community in general.   At the other extreme, such as
the states of Minnesota and Michigan, residents are
vested with a wide array of rights.   Most important,
the public is guaranteed the right to public hearings
concerning the enactment of any environmental
regulation or permit.   They can sue any polluter
violating an environmental law or causing environmental
injury, even if there is no personal injury to health
or property involved.

This disparity of rights runs contrary to an ecosystem
approach to environmental management.   For example,
persons in one ecosystem jurisdiction may have the
right to challenge an environmentally weak water
quality standard and consequently force regulators to
formulate a more appropriate one.   The new standard,
however, may be only marginally beneficial if
governments in upstream jurisdictions do not
correspondingly strengthen their standards and there is
no right for the public to challenge the
appropriateness of such standards.   In this instance,
downstream interests must bear the cost of pollution
from extra-jurisdictional sources.   In the end, the
quality of the entire ecosystem remains at risk.

OVERCOMING THE BARRIERS: THE "ECOSYSTEM RIGHTS ACT"

Reform of the legal impediments to cross-border
litigation within the Great Lakes ecosystem can take
two general routes.   The first route would be to
advocate that each jurisdiction remove those specific
common law and statutory provisions identified as
barriers to cross-border litigation.

The second route is to develop a "model statute" which includes provisions that would vest each resident of the ecosystem with a basic array of environment rights. The rights would include equal access provisions to allow persons to become involved in any legal proceedings within the basin and a set of harmonious rights held by every person to enable them to protect the ecosystem. In effect, this "model statute" would be a code of environment rights for the ecosystem, an "Ecosystem Rights Act" that could be adopted in each basin jurisdiction. Upon adoption by each jurisdiction, uniformity of litigious rights would result, and cross-border litigation would truly be transformed into ecosystem litigation.

The development of a specific regime of environmental rights for Great Lakes (like the Ecosystem Rights Act), can be justified in a number of ways. First, the very fact it is an ecosystem suggests that, due to the interdependence of all elements within the system, an action in one jurisdiction will inevitably affect all others. Hence, the public should have some say in those activities which directly or indirectly affect their interests. Second, the Great Lakes have a special legal status. According to the law of most of the jurisdictions, the Great Lakes is, in effect, a trust property held by the governments for the benefit of present and future generations. This public trust doctrine suggests that every resident of the ecosystem is a beneficiary of that trust, and has certain inherent rights to challenge anyone, including governments, who fails to properly look after the trust property (Olson 1981, chapter 8).

The Ecosystem Rights Act would incorporate the following principles:

> 1. courts and administrative tribunals within ecosystem jurisdictions would consider the ecosystem as a single jurisdictional unit in matters relating to its protection, conservation, and enhancement;

> 2. every person residing within an ecosystem jurisdiction would have equal access to the courts and administrative tribunals within the ecosystem to participate in proceedings relating to the protection, conservation, and enhancement of the ecosystem, and share the same rights to relief as any other person of the jurisdiction where the proceeding was commenced;

3.in any proceeding before a court or tribunal affecting the protection, conservation, or enhancement of the ecosystem, the court or tribunal is empowered to consider the basin-wide impacts of its actions;

4.courts have jurisdictions over any polluter within the ecosystem and will not bar a claim solely because it related to injury to land within the ecosystem but outside the territorial limits of the jurisdiction;

5.every person within the ecosystem has a right to bring action against any person for environmental injury or an order to restrain any activity that may cause harm to the ecosystem or for a violation any environmental law, standard or permit, even though that person does not allege any personal injury;

6.every person within the ecosystem has the right to due notice and the right to make written submissions to the administrative agency or tribunal concerning the enactment any environmental standard or regulation, or the issuance of a permit affecting the ecosystem; and where the public interest demands, the right to a public hearing to challenge the proposed standard, regulation or proposed permit;

7.where an administrative agency or tribunal has made a decision affecting the protection, conservation, or enhancement of the ecosystem, any person within the ecosystem has the right to maintain a court action to challenge that decision, if the public interest so demands.

As with any proposed legislative reform, the success of having the law enacted in each jurisdiction would depend upon the efforts of the support it received. CELRF is publishing a citizen's handbook on this topic as a means to mobilize support for the concept and ensure its relevance and importance is understood by all concerned. Another useful mechanism would be the organization of a governor's campaign, an attempt to convince the decision-makers within each jurisdiction as to the need and importance of the law.

The proposed **Ecosystem Rights Act** is designed to be transferable to any ecosystem with a diversity of

jurisdictions. It essentially attempts to transpose
political borders with ecological borders.

SUMMARY AND CONCLUSIONS

The quest for uniform rights concerning public
participation is only one means to further the
implementation of ecosystem management; a single
dimension in a multi-faceted approach needed to address
the problem of jurisdiction diversity.   Other legal
initiatives must also be researched such as the study
CELRF initiated following the cross-border litigation
study. This more recent examination seeks to determine
the problems associated with the inconsistency of
environmental standards governing toxic pollutants and
the methodologies employed for establishing those
standards. The purpose of the study is to explore the
extent to which uniformity of law throughout the
ecosystem would assist environmental protection goals.

While legal reform cannot unilaterally overcome the
problem of jurisdictional diversity, it does seek to
build one more bridge over very troubled waters.

ACKNOWLEDGEMENTS

The author would like to thank Mr. David Scriven and
Ms. Marcia Valiante for reviewing and providing
insightful comments on drafts of this paper. Gratitude
is also extended to the Joyce Foundation of Chicago and
Environment Canada for their financial support of the
cross-border litigation study.

REFERENCES

Agreement Between the United States of America and
Canada on Great Lakes Water Quality of 1972. 15 April,
1972, T.I.A.S. No. 7312,reprinted in 1972, 11
International Legal Materials 694.

Agreement Between the United States of America and
Canada on Great Lakes Water Quality of 1978, 22
November, 1978, 30 U.T.S. 1383; T.I.A.S. No. 9257.

American Bar Association and Canadian Bar Association, Settlement of International Disputes Between Canada and the U.S.A.   Resolutions adopted by the American Bar Association on 15 August, 1979 and by the Canadian Bar Association on 30 August, 1979 with Accompanying Report and Recommendations. Ottawa: Canadian Bar Association, 20, September 1979 (mimeographed)

Canadian Environmental Law Research Foundation 1985. Jurisdictional Barriers to Environmental Protection in the Great Lakes Basin.   Workshop Proceedings. C.E.L.R.F. Toronto, Ontario.

Chayes, A. 1976.   The role of the judge in public law litigation.   Harv. Law Rev. 89:1281-1316.

Chester, S. 1982.   Private law approaches - Remedies in Canadian courts.   Can. U.S. Law Journal 5:85-90.

Convention on the Protection of the Environment (Nordic Environment).   Reprinted in (1974), 13 International Legal Materials 591.

Great Lakes Water Quality Board 1981.   1981 Report on Great Lakes Water Quality.   Report to the International Joint Commission.   November, Cleveland, Ohio.

International Joint Commission 1982.   First biennial report under the Great Lakes Water Quality Agreement of 1978.   Ottawa, Ontario and Washington, D.C.

Jaffe, L.L. 1968.   The citizen as litigant in public actions: The non-Hohfeldian or ideological plaintiff. Univ. of Penn.   Law REv. 116:1033-1047.

Muldoon,P.R., Scriven, D.A. and Olson, J.M. 1986. Cross-border litigation: legal action in the Great Lakes Ecosystem.   Toronto: Carswell.

Muldoon, P.R. and Stalker, L. 1984.   Equal access: suing polluters on their own turf.   Alternatives 12:12-16.

Munton, D. 1981.   Paradoxes and prospects.   In: The International Joint Commission Seventy Years.   ed. R. Spencer, J. Kirton and K.R. Nossal, Pp. 60-97. Toronto: Centre for International Studies, University of Toronto.

Olson, J.M. 1981.   Michigan Environmental Law. Traverse City, Michigan: Neahtawanta Press.

Organisation for Economic Cooperation and Development 1977. <u>Legal Aspects of Transfrontier Pollution</u>. Paris: O.E.C.D.

Organisation for Economic Cooperation and Development 1974. <u>Problems in Transfrontier Pollution</u>. Paris: O.E.C.D.

Sax, J. 1970. <u>Defending the Environment - A Strategy for Citizen Action</u>. New York: Alfred Knopf.

Treaty relating to boundary waters and questions arising along the boundary between Canada and the United States, (1909). 36 Stat. 2448 (1910), T.S. No. 548.

Uniform Transboundary Pollution Reciprocal Access Act 1983. 9A Uniform Law Annotated 234.

SPECIAL CONTRIBUTION

FROM THE

PEOPLE'S REPUBLIC OF CHINA

CHAPTER 24

LAKE RESOURCES AND FISHERIES UTILIZATION IN HUBEI PROVINCE

Zhang You-Min

Hubei Aquatic Products Science Research Institute, Wuhan, People's Republic of China

There are 636 lakes in Hubei province with a total area of 3000 square kilometers. This number ranks Hubei province first in China. Few areas in the world have such a high concentration of lakes.

REASONS FOR THE FORMATION OF THE LAKES

The lakes are the products of the continuous action of the Yangzhi and the Han Shui river. However, some lakes are formed due to new tectonic shifts.

MORPHOLOGICAL CHARACTERISTICS OF THE LAKES

The lakes are similar to each other in external and internal force conditions. Thus, morphologically there are many common aspects:

1. The lakes are shallow: 520 lakes (93%) have an average depth of 1-4 meters.

2. Plain lake bed and wide littoral zones are typical. The area of the lakes may vary

greatly with high and low water levels (from several to over ten times).

3.Thick silt bottom deposits. Decomposed detritus of aquatic plants and other organics form a soft mud which can be 1-2 meters thick.

While the forms of these lakes are relatively similar to each other, there still exist many differences between them. Specifically differences in geographical characteristics and distribution can be observed.

CHARACTERISTICS OF LAKE HYDROLOGY

Since the natural environment of lakes located in Hubei province is rather simple, the hydrological characteristics of each are almost the same.

1.Lake water levels variations. The annual variations of the water level of lakes are identical with that of the Yangzhi. The changes in lake water levels occurs slightly later. Generally, water levels start to rise in March and April and reach a maximum in August and September. October marks the beginning of the dry season. Around February of the following year, the water level reaches the lowest level. The range of water level change over a year is from one to three meters. Sometimes, it can exceed five meters.

2.Water temperatures are high. Hubei province is located in the semitropics. The temperature is above 20° C during six months of the year, and higher than 15° C for another two months. The highest monthly average water temperature in a year can exceed 30° C and the lowest is usually above 4° C. The phenomenon of freeze-over is rare.

In most lakes the vertical variations of water temperature in summer is within 2° C. For a few lakes with a depth of about 10 meters, the differences of water temperature can exceed 10° C from the water depth of 4 m to 8 or 9 m thus forming an apparent thermocline layer.

3.Water color and turbidity. There is a great diversity of lake water colors, which are light green, green, yellowish green, yellowish brown, and yellow, etc.. This is closely related to the quantity and the size of silt, plankton and other impurities in the water.

4.<u>Chemical properties of the lake waters</u>. A few lakes are adjacent to factories and in suburbs and are polluted. The pollution originates mainly from petroleum, chemical industry, light industry, metallurgy, machinery and electroplating. The major pollutants are organic matter, phenol, chloride, ammonium nitrogen, oil, mercury, chromium (six - valence), arsenic and cadmium, etc. Other pollutants are fertilizers and pesticides. However, the water quality of most lakes is judged to be good.

a.Mineralization

The range of mineralization of lake waters is between 100 - 200 mg/L, hardness is around 3-4°. The lake water is basically soft water, and the water is of the carbonate type.

b.Oxygen

During winter some lakes are supersaturated with oxygen. Generally, the dissolved oxygen in other months is between 80-90% saturation. Even in August, which has the highest water temperature, it can still reach 70%. In certain shallow lakes, which are rich in opiphyte and algae, an oxygen deficient condition will occur in the hypolimnion during extremely hot and windless nights.

c.pH

According to records on 470 lakes, 95% of the lakes have pH values between 6 and 9.

d.Nutrients

Organisms have a rather high nutritional element content (nitrogen, phosphorus, silicon, iron). This is the main reason why the large amount of hydrobiomass exists.

e.Oxygen Demand

The decomposed organic matter content in many lakes is high. Consequently, oxygen consumption by organisms is also high. Sometimes, it exceeds the available dissolved oxygen in the water.

OUTLINE OF TYPES OF ORGANISMS IN LAKES

1.There are more than 80 species of fish in the lake areas. Thirty are the common species, including grass carp, black carp, silver carp, and big head which are peculiar to China.   At present, more than 20 species have been harvested as the main cultured fish.

2.Aquatic plants are abundant in most of the lakes. There are many more than 100 species of aquatic plants. For example, in Hong Lake (area 350 km^2) the average biomass can reach 4.44 kg/m^2 in the fall.   However, in quite a number of lakes, the aquatic plants are not fully utilized.   They are directly transferred into the decomposition and reduction cycle.   This discordant structure between producers and consumers can be solved through the development of fisheries.   Eliminating a portion of the aquatic plants helps to regulate the lakes and prevents them from becoming swamps.

The average range of biomass of phytoplankton in the major lakes is 0.38 - 15.36 mg/L, zooplankton: 0.56 - 3.66 mg/L, benthic animals: 24 - 641 g/m^2.

FISHERY UTILIZATION METHODS AND PRODUCTION LEVELS

Using lakes to develop freshwater fisheries in Hubei Province dates back to the early 1950s.   Three techniques have been used.   The comprehensive management emphasizing fisheries, combining culture with plantation and culture with animal livestock. Each conforms to the concept of large scale comprehensive culture as advocated by the worlds' fish culture specialists.

At the present time, fishing practices associated with a particular lake size can be divided into three types:

1.The majority of lakes are of the middle to large size (area > 30 km^2) and are grouped into the **first** lake type.   Here, emphasis is placed on fishing the natural fish communities.   In order to keep the fishing quantity stable, the following measures are usually taken:

> a.Different lakes and different fishing objectives have their own regulations of definite banned fishing zones, forbidden periods, and regulations restricting harmful fishing tools, methods and catch quantity, etc.

> b.In areas where conditions permit, a small quantity of economical fish and fingerlings can be transplanted and stocked in lakes; or fry are released from the river when the gate

between them is opened at certain periods of the year. This is done in order to replenish fish resources.

In this case, the construction and modification of water structures are not normally undertaken. Fish yield depends on natural resources and therefore is finite.

2. In the **second** type of lakes, the majority are medium-sized (area 7 - 30 km^2). They are part of extensive cultivation with the following characteristics:

a. Management of the lakes is strengthened by raising fry artificially to make maximum use of the natural food in the water bodies and convert it to fish body weight for the purposes of raising fish output.

b. The quantity and kinds of stocked fingerlings must coincide with the food supply capacity of the water bodies. According to results of detailed technical and economic studies, the suitable size of the stock fingerlings is 13.3 cm.. About 5 - 7 kg fish can be harvested from every kg of stocked fingerlings.

c. Besides having regulations for stricter bans of fishing periods and zones, more vigorous and effective measures have been adopted. For example, artificial fish nests have been made to collect eggs.

d. Control of violent predatory fish (mainly false salmon, catfish, menderin fish, mudfish and erythropterus, etc.) Spawning grounds are destroyed and intensive catching is employed.

e. Good blockage equipment prevents the raised fingerlings from escaping.

Thanks to the high productivity of organisms in the lakes, these lakes can also produce a stable and fairly high yield per unit area.

3. The **third** type of lake concerns those where intensive fish culturing as a result of extensive stocking is practiced. Since the natural food in these types of lakes is finite, fish yields can be estimated. For example, a lake with an area of 66 km^2 can provide a fish yield of more than 5 - 10 kg/mu (1 mu = 1/15 ha). Lakes with areas less than 66 km^2 have been observed to

yield no more than 15 - 25 kg/mu. Therefore, to improve fish yield, food and fertilizers must be supplied externally. This activity then becomes an intensive farming operation.

This type of lake is characteristically small in area, usually less than several square kilometers. It is highly controllable. To some extent, it may seem that the pond fish culture has been extended. Not only are sufficient fingerlings stocked, but the quantity of stocking fish can sometimes be several hundred in one mu. Food and fertilizers are also frequently applied. It is usually the combination of feeding and fertilizing that reinforces the activities of the autotrophs and heterotrophs in the water. This reduces the accumulation of substances and enhances the ecological balance in the water. Fish yield is no longer subject to the constraint of limited natural food, but rather to the technical level of intensive farming. The unit area yields are very high. Yields exceeding 200 kg/mu have been measured.

TABLE 1   THE PRODUCTIVE CAPACITY (Level) OF DIFFERENT
          TYPES OF LAKES FOR FISHERIES

TYPES OF FISHERY UTILIZATION	LAKE NAME	AREA (km^2)	RECENT FISH YIELD (kg/mu)
Utilization rate of lakes for fisheries in the whole province of Hubei.		187	22
1st type	Hong Lake	353	7.8
	Lianzi Lake	256	6.3
	Chang Lake	153	6.5
	Futou Lake	128	10.8
2nd type	Wai Dangsun Lake	28	23.8
	Chi Doug Lk.	11.2	17.5
	Ce Lake	7.3	30.0
	Bao Feug Lk.	6.66	27.5
3rd type	Zhu Lin Lake	6	85.0

Examples of the three types of lake fisheries are summarized in Table 1. In the various lake size mentioned here, pen and netcage culture, which is a form of elevated intensive cultivation, has recently emerged. Pen culture, which is similar to a netcage without a bottom, is suitable for application in both shallow water zones and lakes that have a stable water level. The areas required can vary from several to hundreds or even thousands mu. At present the unit area yield is 200 - 400 kg/mu.

Furthermore, it is common to grow economical aquatic plants. Examples are: lotus seeds, lotus roots, and watercultrops in the littoral zones or shallow water zones of the lakes.

LAKE FISHERIES MANAGEMENT AND ENVIRONMENTAL PROTECTION

There are special institutions which are responsible for the management of fishing grounds. They are established for all lakes which have been utilized for fisheries. Responsibilities are as follows:

1.Making rules and regulations for breeding protection and fishing.

2.Producing fry and fingerlings to meet the needs for stocking.

3.Rehabilitation and protection of the fisheries environment.

4.Managing the sale of the product, etc.

At present, the problem of protecting our inland waters has been put on the agenda.

In recent years our country has also issued a series of laws and regulations such as: "Fishery Water Quality Standards", "Regulations of Aquatic Resources Breeding Protection" and "Environmental Protection Law". A national monitoring system of the fishery environment was recently established. The purpose is to strengthen the examination and protection of fishery waters.

CHAPTER 25

PROTECTION OF THE PLATEAU LAKES IN YUNNAN PROVINCE

Liu Fucan, Zhang Xichun and Zhang Jingfang

The Environmental Science Research Institute of Yunnan Province, Kunming, People's Republic of China

INTRODUCTION TO THE PLATEAU LAKES IN YUNNAN

Yunnan Province is located in the southwest area of China, between $21°09'$ - $29°25'$ north latitude and $97°39'$ - $106°12'$ east longitude. It is bounded by Viet Nam and Laos on the south, and adjoins Burma in the east. To the north is the Asian continent and to the south, Yunnan faces the tropical sea. The region belongs to the subtropical climate zone. The provincial area is about four hundred thousand km^2. The altitude for most of the region is 2000 meters above sea level. The general topography is high in the northwest and low in the southeast. The territory slopes from the northwest to southeast like a terraced field. The altitude decreases from 6740 metres to 76 metres above sea level. Since the latitude is low and the altitude varies greatly, Yunnan Province has the characteristic of "stereoscopic topography" and "stereoscopic climate." Therefore, the province is abundant in biological resources and has various types of ecology and land forms.

The annual average temperature of Yunnan is around $15°C$. In the area surrounding the plateau lakes, the seasonal temperature differences are small. The dry and wet seasons are obvious in this area. Summer is not so hot and winter is not so cold. Year round, the weather is spring-like. The annual precipitation is approximately 1000 mm, with 85% of that rainfall concentrated in the rainy season from June to September. Evaporation is about 1200 - 1800 mm with annual sunshine hours between 2000 - 2500 hours and with cumulative temperatures of 4500°C - 6000°C.

More than thirty natural lakes are distributed in various basins between the mountains of Yunnan. The total water area is 1100 km^2 and stored water of thirty billion cubic meters. This represents 6.32% of the total provincial water resources. The main function of the lakes is to supply and regulate water volumes for industrial, agricultural and domestic uses, receiving sewage from cities and tourist attractions, as well as cultivating fish. Most areas around the lake where various nationalities are located, are economically and technologically developed districts.

According to the natural geographical location and distribution, the lakes on Yunnan plateau can be classified into Dianxi and Diandong (west Yunnan and east Yunnan) lake groups.

The Dianxi lake group is distributed along the east border of the Hengduan mountain range. It includes more than ten lakes. The major ones are Erhai Lake, Lugu Lake and Chenghai Lake. The Diandong group contains Dian Lake, Fuxian Lake and Yangzong Lake. Altogether there are more than twenty lakes which are scattered on the Kasite Plateau in the east of Yunnan and east of the Kunming Indentation. Recently, the natural lakes in Yunnan have been changed a great deal. Some of them have been drained for farming or have become marshes. The distribution and characteristics of the major lakes are shown in Tables 1 and 2.

The Yunnan lakes, which are located on the plateaus between 1280 and 3270 meters above sea level, belong to the river systems of Jinsha, Zhu, Hong and Lancang. Most are tectonic lakes formed by fault subsidence. Tectonic lakes also are known as displacement-subsistence lakes. The long axial directions of the these lakes follow the direction of the tectonic line. Most of these lakes occur in a belt which extends in a south-north direction. A great portion of the stratum is limestone or sandstone which has seriously corroded. The corrosion can generally be classified into three types:

1. displacement-subsistence corrosion
2. displacement-subsidence corrosion, and
3. corrosion.

TABLE 1   CHARACTERISTIC STATISTICAL DATA FOR THE MAIN LAKES

LAKE NAME RIVER SYSTEM			Dian Jinsha	Erhai Lancang	Fuxian Nanpan	Yilong Nanpan	Chenghai Jinsha	Lugu Jinsha
DRAIN. BASIN AREA	km^2		2920	2565	759	326	318	248
ANNUAL INFLOW	$x10^8m^3$		7.0	8.17	-	0.85	1.12	-
TOTAL VOLUME	$x10^8m^3$		12.00	28.20	185.0	1.2	27.0	20.72
LAKE AREA	km^2		298.4	250.0	212.0	42.0	78.8	51.8
LAKE LENGTH	km		39.0	40.5	31.5	11.2	20.0	-
LAKE WIDTH	km	max.	12.9	8.4	11.3	-	5.0	-
		min.	2.4	3.4	4.0	-	3.0	-
		ave.	7.66	6.3	7.0	2.2	4.0	2.0
SHORE LENGTH	km		150	117	100	31	45	44
AVE. WATER DEPTH	m		4.1	10.5	8.6	1.97	15.0	40.0
MAX. WATER DEPTH	m		10.0	20.5	155.0	3.7	36.9	90.0
LAKE SURF. ELEV.	m		1885.0	1973.59	1721.22	1411.0	1503.0	2685.0
HIGH. WATER LEVEL	m		1888.29	1975.40	1722.41	-	-	-
LOW. WATER LEVEL	m		1885.15	1973.28	1720.42	-	-	-
AVE. WATER LEVEL	m		1885.0	1973.59	1721.2	1411.0	1503.4	-
SHORE DEV. RATE	-		3.1	2.11	1.81	3.29	-	-

TABLE 2   PHYSICAL-CHEMICAL CHARACTERISTICS OF DIAN, ERHAI'JILU, YILONG, LAKE WATER

PARAMETER		DIAN	ERHAI	JILU	YILONG
pH		8.41 - 9.05	8.35 - 9.05	8.42 - 9.22	8.3 - 8.8
TOT. HARDNESS	(mg/L)	8.4 - 16.9	5.6	10.8 -13.8	4.86
OPACITY	(m)	0.3 - 1.2	4.0	0.3 - 1.2	1.85- 1.9
DO	(mg/L)	2.54 - 6.95	5.64 - 9.33	6.19 - 6.7	7.5 - 9.0
COD	(mg/L)	27.89 - 72.51	1.32 - 3.5	4.02 - 7.08	3.9 - 5.35
BOD	(mg/L)	2.8 - 11.78	0.13 - 2.95	1.25 - 3.69	0.91- 1.36
TOTAL N	(mg/L)	0.58 - 3.19	0.10	0.712- 1.139	0.71 - 0.88
AMMONIA N	(mg/L)	0.00 - 0.45	-	0.212- 1.31	0.04 - 0.22
TOTAL P	(mg/L)	0.048- 0.604	0.011	0.081- 0.099	0.047- 0.07
LAKE TYPE		I*	II*	III*	IV*

*

I   Freshwater shallow lake with medium degree of pollution and eutrophication, formed by downcast of limestone fault in Yun-Gui Plateau subtropical zone.

II   Dystrophic-mesotrophic clean median degree freshwater lake, formed by downcast of inland fault in Yun-Gui subtropical zone.

III   Eutrophic freshwater lake with slight pollution, formed br downcast of limestone fault in Yun-GuiPlateau subtropical zone.

IV   Mesotrophic clean freshwater shallow lake, formed by downcast of limestone fault in Yun-Gui Plateau south subtropical zone.

Most of the rivers that run into the lake are short. Their flow is small and varies seasonally, changing greatly from year to year. Therefore, the water in the lake is mainly supplied by rainfall, resulting in smaller water supplement coefficients for the plateau lakes than are found for those on the plains. The water coefficients for the plateau lakes are not constant and supply is not stable.

The plateau lakes in Yunnan are freshwater lakes with low salt contents and high hardness. The water is of a weak alkaline calcium type. More than thirty lakes were polluted by industrial and agricultural wastewater, as well as domestic sewage. Dian Lake and Jilu Lake have the most heavily polluted lakes and are highly eutrophic. Pollution of the other lakes may be more or less serious, depending on their uses.

The Plateau lakes in Yunnan are rich with aquatic animals and water plants and also exhibit the characteristics of the region. No environmental studies have been conducted on most of these lakes.

The plateau lake group of Yunnan is one of the three large lake groups in China. Many natural freshwater lakes are scattered in the low latitude and high altitude regions. They vary in size and depth and are developed to varying degrees. It offers an opportunity for scientists to play an important part in studying these lakes and the impact of lake environmental pollution and protection.

PROTECTION OF THE YUNNAN PLATEAU LAKES

Lakes are very important to the Yunnan people for living and development of their economy. Historically, the natural ecological characteristic and ecosystem functions of the plateau lakes were not recognized. The relationship between development and protection, short-term and long-term benefits, individual and comprehensive benefits, economic and social benefits, economic and environmental benefits was not considered during the development of the lake resource. Much attention was paid on developing the economy. Environmental protection was neglected. All this resulted in a shortage of water, ecological environmental deterioration, ecological balance dislocation, drought aggravation, calamity increase and obstacles to economic development.

Lakes are natural resources which should be protected as they can be developed more fully and therefore become even more useful. As the lakes are exploited, their comprehensive function should be brought into full play, and their balance and coordination effects upon the ecosystem should be considered. It is necessary to use systems engineering for arranging an integrated relationship between development and protection so that the system can be optimized. As well, economic, social and environmental benefits can be integrated.

To study lake protection from the environmental system optimization point of view, requires most importantly, that water pollution be controlled, water quality be improved and water problems be tackled in a comprehensive way. For example, the serious pollution of Dian lake is influencing the life of the people in Kunming City and hinders its social and economic development and so control of water pollution and comprehensive treatment has become a pressing problem. Some other important work is to protect the ecological environment of the lakes and deal with the ecological problems comprehensively. Every lake is facing some problem which impinges upon all aspects of natural and social ecology. Even though the problems are complicated and extensive, they must be solved.

Our opinions about plateau lake protection in Yunnan are as follows:

Minimum Lake Volume

Fundamentally, a Certain Volume of Water in the Lake Must be Reserved. Most plateau lakes in Yunnan are located in the watershed area which is the source of the rivers. Both catchment area and in-flow to these lakes are small. These lakes are also shallow and located in basins. Thus their volume is limited and surface run-off can not be totally controlled. It is very important to increase the lake volume and regulate capacity. Raising the controlled water level and turning farm land into lake bottom, is one way of increasing the lake volume. At the same time, the economic structure should be changed so that the economic loss caused by flooded land can be compensated. If all of this can be done, a benign ecosystem cycle at the lake region can be maintained and the water resource can be perpetually useful.

Cultivating Forest for Water Reservation

The lakes are located in the basins surrounded by mountains with little forestation. Great quantities of solid are washed away by surface runoff and settle to the lake bottom. The lake becomes shallow and its volume decreases. Consequently, the lake aging process is accelerated. For example, in twenty-six years (1957-1983) the bottom of Dian Lake rose an average of 1.7 cm per year. This means, increasing the mountain green area is basic to improving the lake situation. Growth of trees (especially rapidly growing trees) and grass would protect the soil from washing away. At the same time, cleaning the lake bottom and repairing the lake bank, planting trees for bank protection, as well as constructing settling facilities at the entry of the lake would also be helpful.

Lake Pollution Control

Industry and agriculture are developed at the lake region where the population is concentrated. The lake is the source of the water supply. Most of the lakes in Yunnan are closed or semi-closed lakes and the lake region itself forms an individual ecosystem. As the lake is the lowest datum level for energy flow and material flow, it becomes the place for receiving wastes and is polluted by industrial, agricultural and domestic wastewater. Pollutants such as heavy metals, nutrient elements, etc. seriously affect water quality.

Among the lakes of Yunnan, the Dian Lake is most in need of pollution control. Methods would include:

1. putting the lake under management, limiting new pollution sources,

2. controlling existing pollution sources, discharge pollutant concentration and total discharge volume,

3. reclaiming municipal wastewater for comprehensive utilization to decrease the new water demand, and

4. developing strict discharge standards.

Ecological Initiatives

The readjustment of the ecological structure and the development of the ecological agriculture in the Lake region play a key role in lake protection. The key to this matter is to consider the problem of how the land and water should be used. In the lake region, the most important part of the traditional agricultural economic structure is farming. This should be changed so that agriculture, forestry, stock-raising, industry and other occupations can be developed. Growing rice needs a lot of water. Water consumption for growing industrial crops and vegetable is rather small. Therefore, adjusting the cultivation proportion of these crops appropriately would reduce the pressure for the high demand for water supply.

Forming a man-made ecosystem by developing ecological agriculture could lead to recovery of the ecosystem, which was originally dislocated by an increase in population and the unchecked development of lake resources in the lake region.

Establishing a Protection Area and Developing Tourist Trade

The different types of lakes, various species of animals and plants, abundant resources, special land forms and geological structures in the Yunnan plateau lake region are all valuable objects for scientific study. Moreover the lake regions are scenic places with beautiful weather. Turning the region into a natural protection area, would be helpful in balancing the ecosystem. Scenic spots and historical sites around the lake, interesting local conditions and curious customs of the minorities are suitable ingredients for the development of tourism for the lake region. If this industry could be developed, the income from tourism could support the natural protection area.

Establishing a Mechanism for Lake Management

This concerns establishing a lake management agency to take care of lake control, protection and development, as well as to develop regulations to enhance efficient management of the lake.

STRENGTHENING THE ENVIRONMENTAL PROTECTION STUDY OF LAKES

The plateau lakes in Yunnan play a very important part in the people's life and the national economic fabric. But they have many problems which remain to be solved. Hence major fundamental research combined with a practical study should be started immediately.

The main research work to be done is as follows:

1.Research in Lake Water Resource Management and Protection Strategies, and Their Application.

2.Research in Water Pollution Control Technology and Wastewater Utilization. The focus should be on finding the means to reach the goal.

3.Research on Lake Ecological Environmental Protection Strategies. During the plateau lake resource development in Yunnan, ecological environmental protection is as important as water resource protection. For example, an ecological problem was caused by discharging water for generating power at Erhai Lake. Lugu lake became in part, a marsh, and an environmental geology problem occurred at Yue lake. These phenomena increased the speed of lake aging and degradation processes.

4.Research in Environmental Ecology. The following main points should be determined:

a.environmental ecological characteristics of plateau lakes as study subjects.

b.methods and techniques for studying environmental ecology for the plateau lakes.

c.measures and primary instruments for lake ecology research.

d.technical measures for establishing observation stations at fixed position, i.e. physical-chemical characteristics of the lakes, fixed position observation of growing and declining regularity of aquatic organisms, fixed position observation of

soil erosion and lake basin deposit in lake
regions, fixed observation of lake thermal
conditions and microclimate, etc.

Environmental protection research work of plateau lakes
was started rather late in China.   The foundation is
quite poor and a lot of work has not yet begun.   We
hope to establish scientific and technical cooperation
with all countries in the world so that this new
scientific field may be explored.   Scientists from all
parts of the world are welcome in Yunnan.

CHAPTER 26

THE IMPACT OF LAKE CHANGSHOU ON THE ENVIRONMENT

Zhao Wenqian[1] and Zhu Zhongde[2]

[1] Professor of Hydrological Engineering, Chengdu University of Science and Technology
[2] Senior Engineer, Chengdu Water Power Institute of Surveying and Designing, Chengdu, Sichuan, PRC

ABSTRACT

Lake Changshou is a reservoir constructed in 1957 in Sichuan Province on the Longxi River, a tributary to the Changjiang (Yangtze). It has a surface area of 65 square kilometers and a volume of 1.027 billion cubic meters. The main goal in construction of the reservoir was electrical power generation (total capacity: 104.5 thousand kW). Construction of the reservoir also provided significant benefits to fish and citrus cultivation in the region, as well as smaller but significant benefits to irrigation and navigation. Malarial incidence in the lake region has decreased since the reservoir was filled. Displacement of population and loss of arable land were the main problems encountered after lake construction. Changes in herbaceous and piscine life after filling are discussed in detail. Improvements in waterfowl population are briefly noted. Toxic contamination is not serious, and water quality in the lake is basically good. There is some tendency towards nitrogen eutrophication. Although studies performed after filling indicate that the effects of reservoir construction are strongly beneficial, they also indicate that records for the period prior to construction are less complete than could be wished. For fuller quantification of benefits and environmental

impact, it is suggested that more extensive
investigations be carried out before dam construction.

BACKGROUND

Lake Changshou is located in eastern Sichuan province,
about 100 km from Chongqing. It is a man-made lake
formed by building a dam in 1957 on the Longxi River,
tributary of the Changjiang (Yangtze). Lake Changshou
has a water surface area of 65 km². Its total volume
is 1.027 billion m³. The main function of the lake is
electrical power generation. The district of the lake
has a subtropical, monsoon, moist climate. The weather
is typically cloudy, moist, and extremely hot in
midsummer. Most of the rainfall occurs in early
summer. Spring comes early.

Before formation of the lake, the yearly mean
temperature in the region was 18.5°C; yearly rainfall
1084.3 mm; mean days of annual rainfall 143.6; relative
humidity 90.3 percent; yearly evaporation is 854 mm.
The mean discharge of the river at the location was
44.0 m³/s. Measured maximum flood discharge (1938) was
2,870 m³/s.

The seasonal variation between flood and low flow is
extremely large. More than 60 percent of the annual
runoff of the Longxi River occurs between April and
July. The river has very little sediment. The
measured quantity of sediment at flood time is slightly
higher than 0.1 percent.

Efficiency of land use in the area downstream of the
Longxi River is relatively high. Approximately 90
percent of the total land area is cultivated with
approximately 70 percent of this area is planted with
rice.

IMPACT OF THE FORMATION OF THE LAKE ON SOCIETY AND THE
LOCAL ECONOMY

Improvement of Energy Supply

There are four hydroelectric stations downstream from
the lake, with ten sets of generators and a total
capacity of 104.5 thousand kW. They are important
power stations in the electrical network. They play a

TABLE 1   THE ENVIRONMENTAL IMPACT MATRIX FOR LAKE CHANGSHOU

Matrix	Water Uses (E)	E101 Industrialisation, Commercialisation	E102 Employment	E103 Tourism	E104 Crop and Livestock Farming	E105 Communications	E106 Trade, Local Finance	E109 Recreation	E111 Appearance	E112 Domestic Water Supply	E113 Land Acquisition	E115 Protection against Natural Dangers	E115 Health	E201 Morphology	E203 Suspended Load	E205 Sedimentation	E209 Flooding
A101 Irrigation	3		+1pTMn		+3cPIy	-1cTIn	+2cPMn									k1pPLn	
A102 Energy	1	+3cPHy	+1cPMn		+2cPIy		+3cPIn										
A103 Drinking Water	4									+1cPIy							
A105 Industrial Use	4	+1cPMy															
A106 Navigation	3	<1pPLy	+1pPMn		+2pPMy	+2cPIy	+1cPMn										
A108 Fishing	2	+3cPMy	+1cPMn				+3cPMy										
A201 Presence of Dam						+1cPIy			+2cPIn		-3cPIn	k3pPMy					
A202 Reservoir				k2pPLy		-2cPIy		k1pPLy	k2cPIy								
A204 Construction Site				k2cPIy					k1pPIn								
A208 Quarries and Pits				k1pPLy													
A211 Releases of Water						-1cTMn											
A301 Submerged Areas					-3cPIn	-2cPMn											
A302 Reservoir Surroundings																	
A303 Reservoir Fluctuation Zone					-2cTMy	-1cPMy					-3cPIn		k2pPMn	k1pPMn	-2cPIn	-2cPIn	-3cTIn
A401 Fish Management Restoking +3 PL								k1pPLy							-2cTIn		
A404 Water Level Control												+3pPMy					
A405 Infrastructure												k3pPMy					
A416 Resettlement					k2pPMy												

TABLE 1 continued
THE ENVIRONMENTAL IMPACT MATRIX FOR LAKE CHANGSHOU

Matrix	Water Uses	Impact on Water								Climate	Terrestrial Flora and Aquatic Flora			Terrestrial Fauna and Aquatic Fauna				
		E301 Biology	E302 Physics and Chemistry	E303 Salinity (NaCl, SO₄, etc.)	E304 Solid Loads Turbidity	E305 Temperature	E306 Evaporation	E307 River Flow	E309 Water Table	E401 New Mesoclimate	E504 Cropped Land	E505 Higher Plants	E507 Phytoplankton	E602 Birds	E604 Reptiles and Amphibians	E605 Economic Fish Species	E606 Other Fish Species	E608 Zooplankton
A101 Irrigation	3	xlpPin		xlpPLn			X2pPia	-lcPMn	xlcTin	xlpPMn	xlpTin	+2cPTy		±2pPMn				
A102 Energy	1																	
A103 Drinking Water	4																	
A105 Industrial Use	4																	
A106 Navigation	3													-2cPMn				
A108 Fishing	2					+2cPia		+2cPiy										
A201 Presence of Dam			x3pPMn -3cTin	+1pPLn	+2cPin		-2cPin			x2pPMn		x1pPMn	+2cPMn		x1pPMn	+3cPMy	x1pPMy	+2cPMn
A202 Reservoir																		
A204 Construction Site																		
A208 Quarries and Pits																		
A211 Releases of Water												-1cPin						
A301 Submerged Areas												+3cPMy		+1cPMn				
A302 Reservoir Surroundings												+2cPMn						
A303 Reservoir Fluctuation Zone																		
A401 Fish Management Restoking																+2cPMy	+1cPMy	
A404 Water Level Control																+3cPMy		
A405 Infrastructure																		
A416 Resettlement																		

role in phase frequency and peak load adjustment as well as serving as an alternate source of electricity in the event of accidents, or for safety and economic purposes.

The total electricity generated from 1957 to 1985 is 9.7 billion kWh.   The total monetary return was 0.83 billion Yuan Renminbi with a rate of return on total investment of 4.51 to 1.0.   The cost of electric energy is 0.0086 Yuan Renminbi per kWh.   It is apparent that the benefit of the lake in terms of power generation is significant.

The power generated by the waters of Lake Changshou not only serves the needs of industrial areas in the network, but also has promoted industrial development near the lake district and assisted rural electrification.   After the lake was constructed, a number of pumping stations for irrigation, factories for sugar production, paper mills, and agricultural processing plants were set up.   Local living standards became generally higher.

The Effect of the Lake on Fisheries

Before construction of the lake, the fishing period in the Longxi River was only three months per year and the annual fish catch was 2.5 metric tons.   Since then, a fishery company has been established.   It produces, annually, 11,500 metric tons of fish in addition to 1.3 billion fish fry of a variety of species.   The current average yearly fish catch is 410 metric tons.

With increases in the fish supply and fish catch, a fishery system including the hatching, breeding of fry, the manufacture of fishery machines, and a fishery research institute has gradually evolved.   In the past few years these enterprises made a profit of 0.7-1.0 million Yuan Renminbi per year.   The Aquatic Product Research Institute of Lake Changshou has made significant progress in fishery research.   Its achievements have also had a significant effect on the fisheries in other areas.

Improvement of Navigation

The length of the backwater in the mainstream caused by the dam is 40 km when the lake level is normal.   In tributaries, this increases to 44 km. Navigation is greatly improved, thereby promoting development in the lake district.

Control of Malaria

According to many reports in the literature, lakes and reservoirs often aggravate the spreading of malaria. However, since the formation of Lake Changshou, the malaria incidence has been less than 0.03 percent which is much less than the rate of 6.17 percent prior to formation of the lake.

Because of the presence of the lake, thick growths of grass, conducive to mosquito breeding, have been deeply submerged. Winds prevent a smooth surface water and many mosquito larvae are eaten by fish. As a result, the prevailing base of malaria is weakened. Additionally, improvements in the strong medical and health network in China are important factors for controlling malaria.

The four aspects mentioned are the main benefits of lake formation. The lake also makes valuable contributions to irrigation and flood protection.

Submersion Losses

When established, the lake submerged up to 4,888 hectares of arable land resulting in decreased grain production by 11 thousand tons every year. Forty thousand people had to move.

THE IMPACT OF LAKE CHANGSHOU ON THE NATURAL ENVIRONMENT

1.**After the appearance** of Lake Changshou, the original natural river course was changed into a huge man-made lake with dozens of islands and peninsulas. The lake is located in a red strata, hilly region. The area of the inlake land is about 567 hectares. The bottom of the lake is rough and bumpy, with a mean water depth of 15 m. The maximum depth is more than 40 m.

The hydrological regime characterizing the natural river has changed completely. This lake is operated like a reservoir equalizing flows to downstream users. This results in more economic water   use and better flood control.

2.**The change of hydrological regime** has caused a change in water temperature. Before appearance of the lake the annual mean water temperature of the Longxi River was 20.9°C; monthly mean water temperature varied in the range of 8.4-30.5°C with a monthly range of 22.1°C. Both the rate of water temperature increase decrease are now more rapid.

After the appearance of the lake, the annual mean water temperature is now 17.1°C; the monthly mean water temperature ranges from 10.6-23.4°C, with a monthly range of 12.7°C. This is 9.5°C less than before the appearance of the lake.

It should be noted that the period with water temperatures of 20-28°C   i.e. the most duration for optimum temperatures for cyprinidae (the main family of fish in China) and the organisms that serve as their food, has increased from 4 months to 6 months after appearance of the lake. This significantly affects the growing and breeding of aquatic organisms, as well as the evolution of the lake's population composition.

The water temperature of the former Longxi River was significantly influenced by air temperature as a result of many shallows, steep rapids, and waterfalls. After constructing the reservoir, the rapids have been eliminated, water velocity is greatly reduced, and the water body is deeper. The temperature of water is directly influenced by air temperature only at the surface.   The same factors cause stratification of water temperature.   Thermal stratification occurs mainly from April to October.

In August for example, water temperature in Lake Changshou is highest and stratification is most obvious. The annual mean temperature on the water surface is 31.4°C, on the bottom 14.5°C, a differential of 16.9°C.

The thermocline in the main zone of Lake Changshou, on the has been tentatively determined to be in the depth range of 10-15 m. Above 10 m is the epilimnion; below 15 m is the hypolimnion. As such a simple division is based on relative stabilization.   There are some differences of about 5 m, during the period of thermal stratification, which depend on time and lake conditions.

If the water exchange indices alpha and beta are
calculated on the basis of hydrological data, the
annual mean alpha value is 1.4; the beta value for
flood volume of 3 days is 0.16; and beta for 7 days is
0.21. The maximum alpha was 2.4 (in 1970). If
estimating an extraordinary flood, a whole flood volume
may amount to about 400 million m^3, so that beta is
only about 0.4. Thus, Lake Changshou is a reservoir of
the stable stratification type.

3.**The formation of an impoundment** and its water
temperature can affect the local climate in the lake
region. Variations in air temperature in the lake
region tend to mitigate as a result of constructing the
reservoir. According to statistical analyses with
appropriate correction, the mean the maximum air
temperature in summer decreases about 3°C. The effect
of the lake on the annual mean air temperature is not
noticeable. In addition, the impoundment water
significantly influences the daily air temperature
variation, so that the daily maximum air temperature
occurs at 1300 h at a highland far from the lake shore,
at 1400 h on the lake shore, and at 1500 h in the
centre of the lake. This is characteristic of a
maritime climate.

Statistical analysis also shows that the effect of the
reservoir on the annual mean relative humidity is not
noticeable. However, absolute humidities at different
locations at the same time are quite different. For
example, absolute humidity on Tongxin Island in the
lake and on the lake shore is higher than the absolute
humidity of Changshou Station more than 20 km from the
lake region. It is also higher in summer than in
winter, which is obviously related to evaporation from
the lake surface.

Relative humidity on the lake shore is significantly
higher than at Changshou Station, but the difference
between Tongxin Island and Changshou Station is very
small. Thus, there appears the phenomenon of a higher
relative humidity zone around the lake region. This is
because on the one hand, evaporation on the lake
surface transports water vapor to the lake shore, and
on the other hand, the air temperature around the lake
region decreases more rapidly so as to increase the
relative humidity. Air temperature around the lake
region decreases more rapidly because it is farther
from the water body, which is the heat source, and
because elevation increases gradually with distance
from the lake.

The statistical data after constructing the reservoir
have indicated that the earliest date of first frost in

the lake region was December 18, but at Diangjiang Station, far from the lake, it was November 27, 21 days earlier. The last date of the end of frost in the lake region is February 27, but at Diangjiang it is March 28, 29 days later. The frost period at Diangjiang lasts 72.9 days, but in the lake region it lasts only 26.6 days, a decrease of 46.3 days.

Since the humidity around the lake region is higher than in the adjacent areas, the incidence of fog is higher near the lake. The mean number of foggy days in the lake region is 243.9, but at Changshou Station only 33.75 days. October has the greatest number of foggy days in the lake region, 21.8; March, with the fewest, still has 18.6 days. The number of foggy days in the lake region is more than 200 greater than at Changshou Station. The cause of this can be recognized as the high number of misty days with visibility of more than 100 m in the air on the lake.

4.**The water quality** in Lake Changshou before and after the reservoir construction is obviously different. Formerly, the Longxi River water had no taste and a light green color; (pH = 7.9-8.0; COD 0.8-2.7 mg/L; calcium, magnesium and carbonate contents of 37.2-50.1, 3.7-15.5 and 257 mg/L respectively, chloride content of only 10.4 - 56.1 mg/L. The water quality was judged to be good.

In the period of dam construction, with the increase of construction intensity, although there was no significant change in some basic water quality parameters such as pH and total alkalinity, the total solids, COD, total hardness, and SO_4 parameters increased rapidly, so that the water quality was obviously degraded. Construction activity also decreased the SiO_2 content of the river water, which influenced the productivity of aquatic organisms during certain periods. The above-mentioned parameters were restored gradually after construction. This demonstrates that more attention must be paid to water pollution problems during the construction period.

After formation of the lake, the water appears neutral with slight alkalinity (pH = 7.0-8.4), not considerably different from before. The Ca-Mg ratio is more stable. The regime of dissolved oxygen is good, commonly 6-8 mg/L. The free CO_2 content in the water has decreased. Its mean value in the surface layer is 12.45 mg/L, still in the range of good water quality standards.

As a result of flocculation and settling, Fe^{++}, Fe^{+++} and PO_4^{--} concentrations are very low (0.10-0.038 mg/L). The organic contaminant content (COD) entering

the lake is not high.   After the oxygen-demanding
organic substances enter the lake, ammoniation and
nitrification processes proceed more rapidly.   Because
the assimilating rate of aquatic organisms is high, the
nitrogen content of the water is lower and more stable.
The maximum mean NH_3-N value is only 0.26 mg/L, and
NO_3^--N 0.10-0.071 mg/L.   These values are at the lower
levels of the proper range for fish breeding.

The stratification of the lake also causes differences
in dissolved oxygen and other chemical parameters
between various water layers.   In the bottom layer of
the lake, the dissolved oxygen has been measured at
0.51 mg/L, 8.24% of the upper layer content while the
CO_2 content is 21.12 mg/L, 4.79 times that in the upper
layer.   This is due to the oxygen-consuming deposition
of mud materials.

The hydrochemical reactions caused by stratification of
water temperature and dissolved oxygen makes other
parameters also different between the lake bottom and
surface layers.   They are COD, the three forms of
nitrogen, electrical conductivity, hardness and color.
Considering the three forms of nitrogen for example,
one notes that since generally the surface water is
richer in oxygen than the bottom layer as a result of
wind waves and other dynamic factors, so that at the
surface water appears in an oxidation pattern, the
nitrogen exists mainly in the form of nitrate.   But the
bottom layer has a strong reducing capability with
relatively high contents of ammoniacal and nitrite
nitrogen.

It is possible to improve the water quality in the
reservoir under certain conditions.   At the headwater
of the Longxi River, for example, there are mainly
limestone strata, so that the hardness of the river
water is more than 11^OG, moderately hard.   However,
diluted by river water upstream and downstream in
purple sandstone and shale regions and stored in the
lake, the water becomes softer.

Regarding pollution, prior to entering Lake Changshou
and as a result of the distribution of pollution
sources, in the upper and middle courses, human
contamination of the Longxi River has become more and
more serious at the reservoir entrance.   Thereafter the
pollutant concentration decreases abruptly because of
dilution, and further decrease more gradually towards
the outlet of the reservoir (.e.g. ammoniacal nitrogen,
color, COD) while electrical conductivity, total
hardness and salinity only slightly decreases along the
course of the lake.   This proves that Lake Changshou

actually plays the role of a dilution pond and an oxydizing pond.

As already noted, after creation of the lake, the water temperature became relatively stable. This is favorable to bacteria reproduction. The frequency of exceeding coliform group standards amounts to 0-50%, i.e. 25% on the average. The maximum frequency of exceedence occurs at the drinking water source of the electric power plant. This requires attention.

The transparency of the former Longxi River during low water periods averaged about 150 cm, with a minimum of 100 cm and a maximum of 230 cm. After creation of the lake, the upper range because the current low zone. The mean transparency of low water periods is 102.8 cm, with a maximum of 141.8 cm. When the lake centre and pre-dammed zone become relatively stagnant zones, mean transparency is 217.4 cm. with a maximum of 257.5 cm. The increase in clarity augments the intensity and depth of incident light penetration which is favorable to photosynthetic processes.

In the watershed of Lake Changshou, chemical fertilizer used in the three counties (Changshou, Diangjiang, and Liangping) has been increasing gradually. For example, the mean application level of chemical fertilizers in Liangping County over six years, from 1975 to 1980, was 2.4 kg/ha. Because of the big volume of the lake it is possible to induce eutrophic tendencies. At present, the total phosphorus content of the lake water is 0.003 mg/L; total nitrogen is about 1.06 mg/L. Regarding phosphorus, it is an oligotrophic lake; regarding nitrogen, it is an eutrophic lake. In the months of high water temperature the amount of Cyanophyta is 1.7 times that of Bacillariophyta, i.e. there is a "water bloom" phenomenon in summer. Because industry and mining activities upstream are increasing continuously, there is a significant challenge in trying to prevent such reservoirs from further eutrophication.

EFFECTS OF THE LAKE ON THE BIOTA

1.**After appearance of the lake**, the former ecological system of the river type dramatically evolved to one of the lake type. The feeding organisms and fauna of former river fish have undergone a series of significant changes.

The distinguishing feature of evolution of the phytoplankton in the reservoir was that nitrophilous species (Chlorophyta, Cyanophyta) developed rapidly, prevailing in quantity over those requiring little nitrogen (Chrysophyta, Xanthophyta, Bacillariophyta). For example, in the deep water area near the dam, the ratio of the total amount of Chrysoph Xanthophyta, and Bacillariophyta to that of Chlorophyta and Cyanophyta was 0.23-3.0 in 1958-1959; 0.13-0.47 in 1962-1965; and 0.04 in 1978. Similar tendencies occurred in other areas. The amount of phytoplankton was maintained at approximately 1-3 x 10^6/L.

In the initial stages of reservoir filling, the Rotatoria predominated zooplankton, constituting approximately 90%. The remainder were Protozoa and very few crustaceous animals (Cladocera and Copedoda). After filling this reservoir, planktonic and crustaceous animals developed rapidly as a result of the low water velocity and abundant food, and became the main part of the zooplankton in weight, constituting 60-95% and increasing in total weight. The zooplankton concentration has increased from 1-2 mg/L in the initial stage of reservoir filling to the current 10 mg/L.

The benthic invertebrates of the former Longxi River in the shallow water zones were spiral shells, Lymnaea etc.. After the appearance of the lake, samples of benthos have not been gathered in the gravel stratum area along the river bed. There are some mollusceous animals, such as Anodonta, Lymnaea, etc., and but they are very few. There are more larvae of Chironomus in newly flooded farmland mud. Although the variety and quality of benthic invertebrates has increased after creation of the lake, the total is still small.

There is no record of aquatic flora in the Longxi River before constructing the reservoir, but it has been said that there was no higher aquatic plant life than filamenteous algae. It would be difficult for higher aquatic plants to survive after constructing the reservoir because of the significant increase in water level. Higher aquatic plants occur only in the puddles in the drawdown area: e.g. <u>Potomagaton</u> <u>malainus</u>, <u>P.</u> <u>crispus</u>, <u>Hidrilla</u> <u>verticillata</u>, and <u>Myriophyllum</u> <u>spiratum</u>. But these constitute very little in biomass. However, the runoff during the flood period carries a great number of aquatic floating plants which remain in the reservoir: e.g. <u>Eichhornia</u> <u>crassipes</u>, <u>Azolle</u> <u>imbricata</u>, <u>Salvinian</u> <u>natans</u>, <u>Spirodela</u> <u>polyhiza</u>. These exotic floating plants remain on the water surface of the lake, continuing to produce organic matter through photosynthesis, until they are consumed by the fish.

2.**Before the appearance of the lake**, the fish in the former Longxi River were mainly those which adapted to the ecosystem, such as <u>Varicorhinus</u> <u>angustistomatus</u>, <u>Barbodes</u> <u>sinensis</u>, <u>Sinilabeo</u> <u>rendahli</u>, <u>Procypris</u> <u>Vabaudi</u>, <u>Sinogastromyzon</u> <u>szechuanensis</u> and <u>Hemibagrus</u> <u>macropterus</u>, and those which adapt to any ecosystem, such as <u>Cyprinus</u> <u>carpio</u>, <u>Carassius</u> auratus and <u>Hemiculter</u> <u>leucisculus</u>.

After the appearance of the lake, the water area becomes a good ecological environment to fatten up the <u>Hypopthalmichthys</u> <u>molitrix</u> and <u>Aristichthys</u> <u>nobilis</u>. The backwater area where higher flow velocities prevail also provides beneficial conditions for their spawning and breeding. This is also the result of artificial breeding to reform the native fish fauna. For more than ten years, these two species have made up about 70% of the total fish harvest.

In the former river, fish ate benthic-adhering organisms and omnivorous fish predominated. There are no <u>Hypopthalmichthys</u> <u>molitrix</u> consuming phytoplankton or <u>Aristichthys</u> <u>nobilis</u> swallowing zooplankton. In the initial stage of lake formation, small omnivorous fish increased rapidly. With artificial breeding, plankton consuming <u>H</u>. <u>molitrix</u> and <u>A</u>. <u>nobilis</u> and herbivorous fish such as <u>Ctenopharyngoden</u> <u>idellus</u> are developed as predominant species.

After appearance of the lake, many fish-spawning glutinous eggs (<u>Cyprinus</u> <u>carpio</u>, <u>Carassius</u> <u>auratus</u>, <u>Parasilurus</u> <u>asotus</u>, <u>Parabramis</u> <u>pekinensis</u>, etc.) were obliged to adapt to changed spawning surroundings in different ways because the former spawning surroundings suffered great damage. In spate periods, they spawn eggs that adhere to floating dregs brought by the flood and flooded terrestrial weeds. Their spawning period was frequently affected by flood periods. The fish spawning slightly glutinous eggs, such as Erythroculter and Xenocypris, are more adaptable, and there was no significant effect on their population. As to <u>H</u>. <u>molitrix</u>, <u>A</u>. <u>nobilis</u>, <u>C</u>. <u>idellus</u>, etc., spawning floating eggs, bigger parent fish stocks are formed because of abundant food and good development of sex glands. In the reproduction period, as soon as the water rises upstream, they can naturally spawn on a large scale. However, as the length of high velocity water in the backwater is too short, the majority of eggs still in embryonic stages flow to relatively stagnant water areas and sink to the bottom, thereby meeting a premature end. According to preliminary observations, the survival rate from eggs to fingerling stage is estimated at 40-80 per thousand. Therefore, artificial breeding in Lake Changshou is very important to the development of a fishery.

Knowledge of the experience and succession of predatory fish species in the reservoir is not only of interest for regulating the reservoir fishery, but also a valuable reference for fishery planning of new reservoirs. The predatory fish in the former Longxi River were four species: Opsariichthye uncircostics, Ancherythroculter wangi, P. asotus and H. macropterus. After filling the reservoir, H. macropterus adapted to flowing water surroundings and withdrew to the flowing water areas of the backwater. Their population has drastically decreased. The other three species have developed to varying degrees as a result of improving food and reproduction conditions. The population of O. uncircostris has developed most rapidly. However, the main predatory fish species in the past 25 years have developed continuously with success. This may be summarized as follows: O. uncircostris - A. wangi - Ochetobious elongatus - Erythroculter ilishaeformis. The tendency is development from small to big fish.

3.**The Longxi River** is a relatively small tributary on the upper reaches of the Changjiang River. Before appearance of the lake, and because of the small river surface and small water volume, both the number of species and the amount of waterfowl were few. The construction of the reservoir formed the biggest lake surface in eastern Sichuan and attracted different species and a great number of waterfowl to rest, winter and shelter themselves from enemies. Some islands in Lake Changshou have become breeding sites of Anas poecilorthyncha zonorhyncha.

According to a special count, the waterfowl numbered 15,670 individuals, for which the predominant three species are Anas p. platyrhyncha, A. p. zonorhyncha, and A. C. crecoa. The Ardeidae, Alcedinidae, etc., feeding on fish and shrimp, have also increased. In addition, there are also many phalacrocoraz carbo sinensis, Mergus squamatus, and Larus brunicephalus. The Ardeidae always prey on fry and fingerlings, causing harm to the fishery. They not only consume the energy of the lake ecosystem, but are also the vector of some parasites, spreading fish diseases.

Lake Changshou has a wide drawdown region, which, in addition to its agricultural use, is also an important factor for the reservoir fisher. Every year during low water periods every year, when the drawdown region presents land, a great number of herbaceous plants germinate and grow and become luxuriant vegetation, produce organic matter through photosynthesis, and store energy in the form of plants. When the flood comes and raises the water level, part of the flooded vegetation becomes direct food for fish, and most of it is decomposed by microorganisms into nutrient elements, which are utilized by the phytoplankton and other aquatic plants, thus promoting luxuriant growth of organisms as food for fish.

The change of water and land alternatively in the drawdown region of the reservoir determines the reproduction of the vegetation every year. Some vegetation species in the former Longxi River and on its banks, could not adapt to such drastic changes. The woody and higher aquatic plants died out naturally, while those herbaceous plants adapting to such an environmental change grew rapidly and with great vitality, even to the extent where they predominated.

There are 19 such predominant species such as Rumex maritimus from Polygonaceae, Congza canadensis from Compositae, Hematharia compressa from Gramineae, Fimbristy aesitivalis from Cyuperaceae, and Medicago lupalina, Vicia tetrasperma from Leguminosea, etc. According to estimates from sampling on fixed sites, the vegetative products in the whole drawdown region of the lake amount to 30-34 million kg. Being flooded in water, they provide significant energy to this ecosystem. The vegetation in the drawdown region is characteristically annual wild herbaceous plants, utilized easily by the fish, so that the above-mentioned evolution is very beneficial for fishery production.

Lake Changshou has the effect of a heat source because of its high specific heat. This becomes most obvious during the cold spells in winter. As the water temperature drops under the influence of cold air, the influence of the lake actually raises the lowest air temperature by 0.5-1.0°C. In addition, as this lake has the form of a semi-closed basin, there exists the an added beneficial effect, which raises the mean daily temperature range in the lake region by more than 0.5-1.0°C. This increase is most significant in the low-lying areas surrounding the lake.

As a result of evaporation of lake water and the lake's role as a heat source, the areas adjacent to the lake constitute a relatively higher humidity zone, with mean relative humidity increasing about 3-5%. These features of Lake Changshou have positive impact on the growth and quality of citrus fruits, especially Valencia oranges, a species of the late maturing orange.

Although the mean air temperature of Sichuan Province is low in winter, the extreme lowest air temperature is relatively high and the citrus is seldom subjected to frost injury. This is because of the heating action of the lake waters. Because the air temperature variation at the lake is more gentle, the difference of temperature between days is less, and the interval between the first day with $10^{\circ}C$ and that with $12.8^{\circ}C$ is longer. The growth of the fruit and the accumulation of sugar are very favorable. As investigation has shown, moderate precipitation and high air humidity result in a smooth peel, bright colors, rich juice, sweet taste, special flavor and high yield. As stated, a high relative humidity zone is formed around the lake, which is obviously favorable to growth of Valencia oranges. It should be noted that especially in eastern Sichuan, drought in late summer and followed by a lot of precipitation results in splitting the fruit. However, in the lake region the mean relative humidity significantly increases, from 6-7%, in late summer. The minimum relative humidity is 6-10% and the mean absolute humidity is 2 mb. This all improves the quality of the citrus fruit and prevents splitting fruit.

CONCLUSIONS

The final report of the Committee on Damming and the Environment of the International Commission of Large Dams (ICOLD, "Dams and the Environment," Bulletin 35, June 1980) proposed using a matrix a means of listing and evaluating the impact of individual dams and related construction works on specific parts of the environment. It is hoped it will enable designers and decision makers to take steps to control detrimental effects and accentuate beneficial ones. The environmental impact matrix of Lake Changshou is shown as Table 1. It indicates that reservoirs in the hilly regions situated on the upper part of the Changjiang River Valley have many more beneficial environmental effects than detrimental ones.

Much of the impact of constructing the reservoir is socio-economic. The impact of the natural environment and biota is less.

Of such impacts, the majority are direct, resulting from the lake. As to socio-economic effects, the most important are beneficial effects resulting from electric energy production and fishery improvement. The less important are benefits to irrigation and navigation. Special note should be taken of the flood control aspects.

The main detrimental impacts are land innudation and migration problems. The lake also impacts on the river system, climate and water quality, thereby considerably affecting biota to a considerable degree. The main effects are advantageous conditions for the development of commercial fish and citrus cultivation. On balance then, the environmental impact of the lake is beneficial.

Studies performed after the filling of the reservoir indicated that the environmental background information available before dam construction was incomplete. Thus, although it is clear that the main impact of the reservoir was beneficial, it is difficult to quantify an overall comparison of conditions before and after the appearance of the lake.

It is clear, that more thorough investigations into possible beneficial and detrimental effects would be helpful when constructing reservoirs of this magnitude. Observation, collection, and processing of background information for an environmental impact statement at the design stage will permit prediction and evaluation of beneficial and detrimental effects, and selection of the most advantageous alternative in respect to environmental questions.

Index